THE NATIONAL ACADEMIES
Advisers to the Nation on Science, Engineering, and Medicine

The **National Academy of Sciences** is a private, nonprofit, self-perpetuating society of distinguished scholars engaged in scientific and engineering research, dedicated to the furtherance of science and technology and to their use for the general welfare. Upon the authority of the charter granted to it by the Congress in 1863, the Academy has a mandate that requires it to advise the federal government on scientific and technical matters. Dr. Bruce M. Alberts is president of the National Academy of Sciences.

The **National Academy of Engineering** was established in 1964, under the charter of the National Academy of Sciences, as a parallel organization of outstanding engineers. It is autonomous in its administration and in the selection of its members, sharing with the National Academy of Sciences the responsibility for advising the federal government. The National Academy of Engineering also sponsors engineering programs aimed at meeting national needs, encourages education and research, and recognizes the superior achievements of engineers. Dr. Wm. A. Wulf is president of the National Academy of Engineering.

The **Institute of Medicine** was established in 1970 by the National Academy of Sciences to secure the services of eminent members of appropriate professions in the examination of policy matters pertaining to the health of the public. The Institute acts under the responsibility given to the National Academy of Sciences by its congressional charter to be an adviser to the federal government and, upon its own initiative, to identify issues of medical care, research, and education. Dr. Harvey V. Fineberg is president of the Institute of Medicine.

The **National Research Council** was organized by the National Academy of Sciences in 1916 to associate the broad community of science and technology with the Academy's purposes of furthering knowledge and advising the federal government. Functioning in accordance with general policies determined by the Academy, the Council has become the principal operating agency of both the National Academy of Sciences and the National Academy of Engineering in providing services to the government, the public, and the scientific and engineering communities. The Council is administered jointly by both Academies and the Institute of Medicine. Dr. Bruce M. Alberts and Dr. Wm. A. Wulf are chair and vice chair, respectively, of the National Research Council.

www.national-academies.org

D1018596

OCEAN NOISE AND MARINE MAMMALS

Committee on Potential Impacts of Ambient Noise
in the Ocean on Marine Mammals

Ocean Studies Board

Division on Earth and Life Studies

NATIONAL RESEARCH COUNCIL
OF THE NATIONAL ACADEMIES

THE NATIONAL ACADEMIES PRESS
Washington, D.C.
www.nap.edu

THE NATIONAL ACADEMIES PRESS   500 Fifth Street, N.W.   Washington, DC 20001

NOTICE: The project that is the subject of this report was approved by the Governing Board of the National Research Council, whose members are drawn from the councils of the National Academy of Sciences, the National Academy of Engineering, and the Institute of Medicine. The members of the committee responsible for the report were chosen for their special competences and with regard for appropriate balance.

This report and the committee were supported by the National Oceanographic Partnership Program with funds provided by the Office of Naval Research, the National Oceanic and Atmospheric Administration, the National Science Foundation, and the U.S. Geological Survey. The views expressed herein are those of the authors and do not necessarily reflect the views of the sponsors.

International Standard Book Number 0-309-08536-5 (Book)
International Standard Book Number 0-309-50694-8 (PDF)

Library of Congress Control Number 2003103681

Additional copies of this report are available from the National Academies Press, 500 Fifth Street, N.W., Lockbox 285, Washington, DC 20055; (800) 624-6242 or (202) 334-3313 (in the Washington metropolitan area); Internet, http://www.nap.edu

Preface

The Committee on Potential Impacts of Ambient Noise in the Ocean on Marine Mammals was charged with assessing our state of knowledge of underwater noise and recommending research areas to assist in determining whether noise in the ocean adversely affects marine mammals. The committee was selected to represent a diverse range of expertise, including acousticians and marine biologists, as well as an expert in geophysical exploration. The committee convened four times, beginning in March of 2001, including three open public sessions. A wide variety of experts in the field of marine mammals and noise addressed the committee and submitted materials for review. The committee is indebted to the following for their assistance: Dan Costa, University of California, Santa Cruz; Jim Finneran, Space and Naval Warfare (SPAWAR) Systems Center; Charles Greene, Greeneridge Sciences, Inc.; Richard Heitmeyer, Naval Research Lab; David Kastak, University of California, Santa Cruz; Charles Liberman, Harvard University; Bertl Møhl, Aarhus University; Paul Nachtigall, Hawaii Institute of Marine Biology; Charles O'Neill, Naval Oceanographic Office; Sam Ridgway, SPAWAR Systems Center; Ron Schusterman, University of California, Santa Cruz; Peter Tyack, Woods Hole Oceanographic Institution; and William Watkins, Woods Hole Oceanographic Institution.

In addition, valuable input to the committee's work was provided by a number of researchers. The committee would like to offer sincere appreciation to Douglas Cato, Australia Department of Defence; Elena McCarthy, University of Rhode Island; Jennifer Miksis, University of Rhode Island; Kevin Smith, Naval Postgraduate School; and Eryn Wezensky, University of Rhode Island.

Two previous National Research Council reports examined the possible consequences of ocean noise on marine mammals. *Low-Frequency Sound and Marine Mammals: Current Knowledge and Research Needs* (NRC, 1994) provided an initial survey of our understanding of the impacts of marine noise on mammals. The second report, *Marine Mammals and Low-Frequency Sound: Progress Since* 1994 (NRC, 2000), primarily reviewed the marine mammal research conducted as part of the Acoustic Thermometry of Ocean Climate (ATOC) experiments. Both reports provided a suite of recommendations, many of which still apply and some of which will be reiterated in this report.

Coincident with the deliberations of this committee, two Navy sonar systems received a very high level of attention from the press and environmental organizations. Use of one of the Navy sonar systems, the SQS-53C, a mid-range active sonar, was found to contribute to a stranding incident in the Bahamas. In addition, a separate, low-frequency active sonar system, the SURTASS-LFA, was approved by the National Marine Fisheries Service for use by the U.S. Navy. Both of these systems are discussed in this report, since they contribute noise to the oceans, but neither is discussed in detail.

One of the challenges in preparing this report was to standardize the units of measure. Another was to clarify commonly used terms in underwater acoustics, seismic exploration, and marine mammology. A glossary is included to assist with some of the general terminology in the report.

Chapter 1 of this report provides a brief overview of the issues pertaining to marine mammals and noise and the committee's approach to answering its charge. Introductory material describes the physics of underwater sound, as a rudimentary understanding of these principles is necessary to understand the material that follows. Chapter 2 describes both natural and human contributions to noise in the ocean and discusses long-term trends in noise levels. Chapter 3 describes effects of ocean noise on marine mammals, focusing primarily on behavioral changes. Models of marine sound and its effects on marine mammals are described in Chapter 4. Chapter 5 contains findings and recommendations of the committee, drawing on the content of the previous chapters.

Acknowledgments

This report has been reviewed in draft form by individuals chosen for their diverse perspectives and technical expertise, in accordance with procedures approved by the NRC's Report Review Committee. The purpose of this independent review is to provide candid and critical comments that will assist the institution in making its published report as sound as possible and to ensure that the report meets institutional standards for objectivity, evidence, and responsiveness to the study charge. The review comments and draft manuscript remain confidential to protect the integrity of the deliberative process. We wish to thank the following individuals for their review of this report:

Whitlow Au, University of Hawaii, Kaneohe
Douglas Cato, Department of Defence, Canberra, Australia
Robert Hofman (ret.), Marine Mammal Commission, Bethesda, Maryland
Elena McCarthy, University of Rhode Island, Newport
John Potter, National University of Singapore
Henrik Schmidt, Massachusetts Institute of Technology, Cambridge
Jörn Thiede, Alfred Wegner Institute, Bremerhaven, Germany
Peter Tyack, Woods Hole Oceanographic Institution, Massachusetts

Although the reviewers listed above have provided many constructive comments and suggestions, they were not asked to endorse the conclusions or recommendations nor did they see the final draft of the report before its release. The review of this report was overseen by **Robert Knox,** Scripps

Institution of Oceanography. Appointed by the National Research Council, he was responsible for making certain that an independent examination of this report was carried out in accordance with institutional procedures and that all review comments were carefully considered. Responsibility for the final content of this report rests entirely with the authoring committee and the institution.

Contents

Executive Summary

In recent years, both the scientific community and the general public have become increasingly aware of—and concerned about—conserving the earth's marine resources. Heightened concerns are evident from the increase of scientific and popular articles devoted to such topics as beach closures, harmful algal blooms, and marine mammal strandings. Among the most sensitive and controversial yet least understood subjects is the effect of human-generated noise on marine mammals. Scientists and laypersons alike are well aware that human-generated sound in the sea comes from a variety of sources, including commercial ship traffic, oil exploration and production, construction, acoustic research, and sonar use. Underwater sounds are also generated by natural occurrences such as wind-generated waves, earthquakes, rainfall, and marine animals. It is well known that noise levels in the sea began to increase steadily with the onset of industrialization in the mid-nineteenth century. The conventional assumption is that this trend has continued in recent times as well, but there is only limited scientific evidence to support this hypothesis. Many factors have combined to escalate the awareness of and concern for noise impacts on marine mammals and on their habitat, supporting communication systems, and behavior. However, remarkably few details are known about the characteristics of ocean noise, whether it be of human or natural origin, and much less is understood of the impact of noise on the short- and long-term well-being of marine mammals and the ecosystems on which they depend.

It was in this context of these uncertainties that the current committee effort began. At the request of the federal interagency National Ocean Partnership Program, with sponsorship from the Office of Naval Research, the National Oceanic and Atmospheric Administration, the National Science Foundation, and the U.S. Geological Survey, the National Research Council (NRC) of the National Academies undertook a study to examine the current state of knowledge on ocean noise and its effects on marine mammals. The NRC was asked to

- evaluate the human and natural contributions to marine ambient noise and describe the long-term trends in ambient noise levels, especially from human activities;
- outline the research needed to evaluate the impacts of ambient noise from various sources (natural, commercial, naval, and acoustic-based ocean research) on marine mammal species, especially in biologically sensitive areas;
- review and identify gaps in existing marine noise databases; and
- recommend research needed to develop a model of ocean noise that incorporates temporal, spatial, and frequency-dependent variables (Box 1).

The committee held three public meetings and received input from underwater acousticians, marine mammalogists, auditory physiologists, and naval oceanographers. The committee reviewed previous NRC reports (NRC, 1994, 2000), current scientific articles, symposium reports, models, and data compiled by the Naval Oceanographic Office.

This report is the third in a series by the NRC examining the potential effects of ocean noise on marine mammals. Although the three reports evolved from very different charges and were generated by separate committees, many similar research needs became evident during each study. The recommendations presented in this report build on, but do not replace, those presented in the earlier efforts (NRC, 1994, 2000). This committee recommends that all three reports be examined in order to better understand the research needs required to mitigate the effects of human-generated ocean noise on the marine ecosystem.

FINDINGS

For the purposes of evaluating the potential effects of underwater sound on the marine environment, both ambient noise and noise from identifiable sources must be considered. The term "ambient noise" is used by the underwater acoustics community to refer to the background din emanating from a myriad of unidentified sources. Although the type of noise source may be known, the specific sources are not identified. When examining the possible effects of ocean noise on marine mammals noise from specific sources is also important; therefore, the term "ocean noise" will be used in this report to refer to all types of noise sources.

Sound in the ocean is generated by a broad range of sources, both natural and human (anthropogenic), for intentional use or as the unintended consequence of activity in the ocean. Natural geophysical sources include wind-generated waves, earthquakes, precipitation, and cracking ice. Natural biological sounds include whale songs, dolphin clicks, and fish vocalizations. Anthropogenic sounds are generated by a variety of activities, including commercial shipping, geophysical surveys, oil drilling and production, dredging and construction, sonar systems, and oceanographic research. Intentional sounds are produced for an explicit purpose, such as seismic surveying to find new fossil fuel reservoirs. Unintentional sounds are generated as a byproduct of some other activity, such as noise radiated by a ship's machinery as it crosses the ocean.

A proper accounting of the global ocean noise budget must include both the background ambient component and the contributions from identifiable sources. An overall global noise budget typically is derived by averaging the received noise spectrum over space and time. Contributions from transient-in-time and localized-in-space components are lost in this averaging process. This conventional accounting technique suggests that the two largest contributors to the overall (space- and time-averaged) deep-ocean noise budget are wind-generated ocean waves over the frequency band from 1 Hz to at least 100 kHz and commercial shipping at low frequencies (from 5 Hz to a few hundred Hz). However, it is clear also that this method is only one approach to computing the noise budget and is not necessarily the most appropriate one for assessing the impact of sound on marine mammals.

There are very limited data to determine long-term trends in ocean noise levels. While noise levels in the ocean began to increase with the onset of the Industrial Revolution (ca. 1850), it is much less clear that this trend is continuing in the twenty-first century. Commercial shipping noise is actually the only area for which educated speculation on long-term trends is possible. On one hand, the substantial increase in the number of commercial vessels during the past 50 years, supplemented by limited noise observations, implies there has been a gradual increase in noise levels from ship traffic on the order of 15 dB. On the other hand, newer ships may be quieter, and the relationship between ship-radiated noise and ship parameters (e.g., gross tonnage, length, and speed) is not sufficiently understood to develop a reliable predictive capability. Although evidence on long-term trends in ocean noise characteristics is very limited and there is even less evidence on the effects of ocean noise on marine life, present data are sufficient to warrant increased research and attention to trends in ocean noise.

There are very limited observations concerning the effects of ocean noise on marine mammals. Short- and long-term effects on marine mammals of ambient and identifiable components of ocean noise are poorly

Box 1
Overview of the Committee's Research Recommendations

To Evaluate Human and Natural Contributions to Ocean Noise

- Gather together in one location existing data on man-made sources and noise;
- Measure alternative properties of man-made sources in addition to average acoustic pressure spectral level;
- Establish a long-term ocean noise monitoring program covering the frequency band from 1 to 200,000 Hz;
- Monitor ocean noise in geographically diverse areas with emphasis on marine mammal habitats;
- Develop quantitative relationships between man-made noise and levels of human activity;
- Conduct research on the distribution and characteristics of marine mammal sounds;
- Develop a global ocean noise budget that includes both ambient and transient events and uses "currencies" different from average pressure spectral levels to make the budget more relevant to marine mammals.

To Describe Long-Term Trends in Ocean Noise Levels, Especially from Human Activities

- Establish a long-term ocean noise monitoring program covering the frequency band from 1 to 200,000 Hz;
- Develop quantitative relationships between man-made noise and levels of human activity.

Research Needed to Evaluate the Impacts of Ocean Noise from Various Sources on Marine Mammal Species

- Measure effects of alternative properties of man-made sources in addition to average acoustic pressure spectral level on marine mammals;
- Establish a long-term ocean noise monitoring program covering the frequency band from 1 to 200,000 Hz;
- Monitor ocean noise in geographically diverse areas with emphasis on marine mammal habitats;
- Try to structure all research on marine mammals to allow predictions of population-level consequences;
- Identify marine mammal distributions globally;
- Conduct research on the distribution and characteristics of marine mammal sounds;
- Develop short-term, high-resolution, and long-term tracking tagging technologies;

- Search for subtle changes in behavior resulting from masking;
- Search for noise-induced stress indicators;
- Examine the impact of ocean noise on nonmammalian species in the marine ecosystem;
- Continue integrated modeling efforts of noise effects on hearing and behavior;
- Develop a marine-mammal-relevant global ocean noise budget;
- Investigate the causal mechanisms for mass strandings and observed traumas of beaked whales.

Current Gaps in Existing Ocean Noise Databases

- Gather together in one location existing data on man-made sources and noise;
- Measure alternative properties of man-made sources in addition to average acoustic pressure spectral level;
- Establish a long-term ocean noise monitoring program covering the frequency band from 1 to 200,000 Hz and which includes transients;
- Monitor ocean noise in geographically diverse areas with emphasis on marine mammal habitats;
- Conduct research on the distribution and characteristics of marine mammal sounds.

To Develop a Model of Ocean Noise that Incorporates Temporal, Spatial, and Frequency-Dependent Variables

- Gather together in one location existing data on man-made sources and noise;
- Measure alternative properties of man-made sources in addition to average acoustic pressure spectral level;
- Establish a long-term ocean noise monitoring program covering the frequency band from 1 to 200,000 Hz (data are critical for model validation);
- Monitor ocean noise in geographically diverse areas with emphasis on marine mammal habitats;
- Develop quantitative relationships between man-made noise and levels of human activity;
- Conduct research on the distribution and characteristics of marine mammal sounds;
- Incorporate distributed sources into noise effects models;
- Develop a marine-mammal-relevant global ocean noise budget.

Administrative Recommendations

- Provide a mandate to a single federal agency to coordinate ocean noise monitoring and research, and research on effects of noise on the marine ecosystem;
- Educate the public.

understood. There is no documented evidence of ocean noise being the direct physiological agent of marine mammal death under any circumstances. On the other hand, marine mammals have been shown to change their vocalization patterns in the presence of background and anthropogenic noise. Furthermore, the long-term effects of ambient noise on marine organisms are even less well understood. Potential effects include changes in hearing sensitivity and behavioral patterns, as well as acoustically induced stress and impacts on the marine ecosystem.

Models describing ocean noise are better developed than models describing marine mammal distribution, hearing, and behavior. The biggest challenge lies in integrating the two types of models. A wide variety of ambient noise models and databases have been developed by the U.S. Navy as part of its antisubmarine warfare effort. However, the focus on naval scenarios means that they are not ideally suited for marine mammal applications. Models of marine mammal habitats and distribution patterns, as well as effects models linking dosage and response, are severely limited by a paucity of data. To provide a product that is useful for understanding and managing interactions between marine mammals and noise, existing databases must be expanded, updated, and coordinated to allow the integration of both marine mammal and ocean noise models. Well-documented databases also are essential for performing the critical step of model validation.

Recent reports both in the press and from federal and scientific sources indicate that there is an association between the use of high-energy mid-range sonars and some mass strandings of beaked whales. Recent mass strandings of beaked whales have occurred in close association, both in terms of timing and location, with military exercises employing multiple high-energy, mid-frequency (1-10 kHz) sonars. In addition, a review of earlier beaked whale strandings further reinforces the expectation that there is at least an indirect relationship between the strandings and the use of multiple mid-range sonars in military exercises in some nearshore beaked whale habitats. Several press reports about the recent incidents appeared while this report was in preparation and attributed the strandings to "acoustic trauma." Acoustic trauma is a very explicit form of injury. In the beaked whale cases to date, the traumas that were observed can result from many causes, both directly and indirectly associated with sound, but similar traumas have been observed in terrestrial mammals under circumstances having no relation to sound exposure. Careful sampling and analysis of whole animals have rarely been possible in the beaked whale cases so far, which has made definitive diagnoses problematic. As of this writing, eight specimens in relatively fresh condition have been rigorously analyzed. Because of the repeated associations in time and location of the strandings and sonar in military exercises, the correlation between sonars and the strandings is compelling, but that association is not synonymous with a causal mechanism for the deaths of the stranded animals. The cause of death in all cases was attributed to hyperthermia, but a precise cause for the unusual traumas that were also seen in the cases examined has not yet been determined. The NATO/SACLANT Undersea Research Center report (D'Amico and Verboom, 1998) and the joint NOAA-Navy interim report (Evans and England, 2001) have not been discussed in detail in this document because of the preliminary nature of the findings. However, this is clearly a subject needing much additional research. The research program outlined in Evans and England is a good start.

RECOMMENDATIONS

A federal agency should be mandated to investigate and monitor marine noise and the possible long-term effects on marine life by serving as a sponsor for research on ocean noise, the effects of noise on marine mammals, and long-term trends in ocean noise. Federal leadership is needed to (1) monitor ocean noise, especially in areas with resident marine mammal populations; (2) collect and analyze existing databases of marine activity; and (3) coordinate research efforts to determine long-term trends in marine noise and the possible consequences for marine life.

Existing data on marine noise from anthropogenic sources should be collected, centralized, organized, and analyzed to provide a reference database, to establish the limitations of research to date, and to better understand noise in the ocean. Currently, data regarding noise produced by shipping, seismic surveying, oil and gas production, marine and coastal construction, and other marine activities are either not known or are difficult to analyze because they are maintained by separate organizations such as industry database companies, shipping industry groups, and military organizations. It would be advantageous to have all data in a single database in order to improve the ability of interested parties to access the data sets and use them in research, for scientific publications, in education, and for management and regulatory purposes. This database could be a distributed network of linked databases, using a standardized series of units of measure. International cooperation in this database development effort as well as international access to the information should be encouraged, since the marine mammal and ocean noise issue is global.

Acoustic signal characteristics of anthropogenic sources (such as frequency content, rise time, pressure and particle velocity time series, zero-to-peak and peak-to-peak amplitude, mean squared amplitude, duration, integral of mean squared amplitude over duration, repetition rate) should be fully reported. Each characteristic of noise from anthropogenic sources may differentially impact each species of marine mammals. The complex interactions of sound with marine life are not sufficiently understood to specify which features of the acoustic signal are important for specific impacts. Therefore as many characteristics as possible should be measured and reported. For transients, publication of actual acoustic pressure time series would be useful. Experiments that expose marine mammals to variations in these characteristics should be conducted in order to determine the physiological and behavioral responses to different characteristics. Particular attention should be paid to the sources that are likely to be the large contributors to ocean noise in especially significant geographical areas and to sources suspected of having significant impacts on marine life.

A long-term ocean noise monitoring program over a broad frequency range (1 Hz to 200 kHz) should be initiated. Monitoring and data analysis should include average or steady-state ambient noise as well as identifiable sounds such as seismic surveying sources, sonars, and explosive noises that are not identified in classical ambient noise data sets. Acoustic data collection should be incorporated into global ocean observing systems initiated and under discussion in the United States and elsewhere. A research program that develops a predictive model of long-term noise trends should be initiated. Data from monitoring systems should be available in a timely manner to facilitate informed decision making by interested industry, military, and marine researchers, operators, and regulatory agencies.

Efforts to measure ocean noise should be targeted toward important marine mammal habitats. Until these habitats are fully described, it is reasonable to begin a long-term monitoring program in coastal areas, locations close to known marine mammal migration paths, foraging areas, and breeding grounds. As new marine mammal habitats are identified, these should be added to the acoustic surveys in order to provide a complete picture of the acoustic environment in important marine mammal ecosystems.

A research program should be instituted to investigate the possible causal relationships between the ambient and identifiable source components of ocean noise and their short- and long-term effects on marine organisms. Addressing this challenging and difficult problem will require a multidisciplinary effort between biologists and acousticians to establish a rigorous observational, theoretical, and modeling program. An initial significant focus of this work should be the examination of the possible relationship between the acoustics of identifiable high-energy, mid-frequency sonars, marine mammal trauma, and mass stranding events. In addition, a study of the potential influence of ambient noise on long-term animal behavior should be vigorously pursued.

Whenever possible, all research conducted on marine mammals should be structured to allow predictions of whether responses observed indicate population-level effects. Although it is difficult to obtain direct evidence of impacts of human activity on marine mammals, it is even more difficult to determine long-term impacts on individuals or impacts on populations. Although the few documented cases of direct impact on individuals have raised awareness of potential population impacts, no measures exist of marine mammal population effects from ocean noise.

Research should be conducted beyond locales already known and studied to globally characterize marine mammal distributions and populations. Despite the large body of marine mammal research to date, including what was recommended in previous reports (e.g., NRC, 1994), there is a surprising lack of information regarding the global distribution of marine mammals. Migration routes, breeding grounds, and feeding areas are known for relatively few species. In order to predict the importance of noise effects on marine mammal behavior, the seasonal and geographic distribution of the mammals must be better known both through survey data and through the use of predictive oceanographic variables, such as topography, bottom type, and water column variables. This enormous task will require the development of new sampling and extrapolation techniques in order to be practically achievable.

Research to determine quantitative relationships between levels of anthropogenic activity and noise should be conducted. For example, if there is a robust relationship between vessel type and noise, vessel traffic data could be used to predict shipping noise. Identifying reliable indicators for anthropogenic sources will provide an additional modeling tool and predictive capability that will be particularly useful in areas where long-term monitoring may be difficult or impossible. Similar needs exist for every facet of human activity in the oceans.

Research should be undertaken to describe the distribution and characteristics of sounds generated by marine mammals and other marine organisms seasonally, geographically, and within behavioral contexts. While good progress has been made in describing marine mammal acoustic repertoires, much less is known about the details of natural patterns of sound production, including the means of production and context in which different vocalizations are produced, as well as how they vary diurnally, seasonally, and geographically. Marine mammals themselves may be significant sources of ocean noise, although possibly in localized areas over limited time periods. These studies will also shed light on the contribution that marine organisms make to the global ocean noise budget.

Research should be conducted to determine subtle changes in marine mammal behavior, as well as failure to detect calls from other animals or echoes from their own echolocation, that might result from masking of biologically important acoustic information by anthropogenic sounds. Short-term responses of marine mammals to anthropogenic noise sources have been documented to a limited degree; however, long-term effects of marine noise on the behavior of marine mammals have received less attention. Impacts resulting from increases in background ambient noise have not been documented.

   Marine mammal tagging studies should be continued to observe behavioral changes in response to acoustic cues and to provide important data for simulation models. Efforts to improve marine mammal tagging technology should continue to receive support. Two technological improvements of current tags are needed: (1) increase the duration of long-term data gathering tags from months to multiple years to observe annual behavior cycles and migration patterns, and (2) extend the duration of high-resolution tags from hours to days to gather more data on daily behavior and environmental cues. Current tagging technology allows individual marine mammals to be tracked up to months. Tags capable of higher-resolution data collection, including animal orientation, acceleration, and produced or received sounds, can generally collect data for less than one day. These data have proven very valuable in determining behavioral patterns in a variety of cetaceans and pinnipeds and correlating their behavior with environmental cues. The technology should continue to be developed to allow longer studies using both the high- and low-resolution tags.

   Research efforts should seek to determine if reliable long-term stress indicators exist and if they can be used to differentiate between noise-induced stress and other sources of stress in representative marine mammal species. Stress indicators may be one useful marker for long-term effects of anthropogenic noise on marine mammals.

   The impact of noise on nonmammalian organisms in the marine ecosystem should be examined. Fish use sound in many ways that are comparable to the ways marine mammals communicate and sense their environment. The effects of anthropogenic noise on fish and other nonmammalian species, including their eggs and larvae, are largely unknown. As cohabitants of the marine ecosystem and as members of the same food web, noise impacts on marine fish could, in turn, affect marine mammals.

   Modeling efforts that integrate acoustic sources, propagation, and marine mammals should be continued and fully supported. Simulation models that predict the characteristics of the noise (frequency content, mean squared level, peak level, pressure time series, etc.) and their effects on marine mammals may assist in understanding and mitigating harmful effects of marine noise on mammals. At least one such effort is underway: the Effects of Sound on the Marine Environment model sponsored by the Office of Naval Research. Modeling some direct physiological effects on hearing (e.g., temporary or permanent threshold shift) is relatively straightforward, although limited by the small data sets available from a limited number of species. These integrative tools should be expanded to include the effects of sources of noise that may change their distribution over time such as shipping, wind-induced breaking waves, and distributed biological noise. More effort should be placed on modeling, both explicit marine species hearing models and behavioral effects models for all types of ocean noise.

   A model of global ocean noise that properly reflects the impact of both ambient noise and noise from identified sources on marine mammals should be developed and verified. The conventional approach that utilizes an average pressure spectrum budget is limited in its application to the marine mammal problem. A more comprehensive approach that encompasses contributions of both transient events and continuous sources to ocean noise should be pursued. Many of this committee's recommendations, particularly those concerning information on distribution and source signatures of man-made sources, must be addressed in order to have the capability to develop a marine-mammal-relevant global ocean noise model. In addition, since model validation is a critical part of the model development process, the committee's recommendations pertaining to the collection of high quality, well-documented ocean noise data sets must be pursued in tandem.

A program should be instituted to investigate carefully the causal mechanisms that may explain the traumas observed in beaked whales, whether this is a species-specific or broader issue, and how the acoustics of high-energy, mid-range sonars may directly or indirectly relate to mass stranding events. The research program outlined in Evans and England (2001) represents a good initial effort. The association of beaked whale mass strandings with high-energy, mid-range sonars has recently received much public attention, and the preliminary scientific findings of two such events have been released in agency reports but have not appeared in the peer-reviewed literature. Review of prior mass stranding reports for beaked whales further reinforces the probability of this relationship. In few cases have the beaked whale carcasses been in a condition to allow full, definitive forensic analyses. The complexity of obtaining appropriate samples from stranded beaked whales and the paucity of data to date, both from mass and nonmass strandings, prevent clearly determining the mechanisms and any causal relationship behind the traumas observed, the strandings per se, and sonar use.

The committee encourages the acoustical oceanography community, marine mammal biologists, marine bioacousticians, and other users of sound in the ocean, such as the military and oil industry, to make greater efforts to raise public awareness of fundamental acoustic concepts in marine biology and ocean science so that they are better able to understand the problems, the need for research, and the considerable potential for solving noise problems. The public, including environmental advocates, are very interested in anthropogenic noise in the ocean and its effect on marine animals. Recently there has been a communication gap between users of sound in the ocean, including scientists, and the public. Much of the gap in understanding between the ocean science community and the public arises from the public's lack of understanding of fundamental acoustic concepts and the scientific community's failure to communicate these concepts effectively. Source and received levels, propagation loss, air-water physical acoustic differences, and the term "decibel" are examples of concepts that have been misunderstood by the media, environmental organizations, and the general public.

1

Introduction

The environment, whether in sea or on land, is filled with natural sounds, although increasingly many locales have sound contributed by anthropogenic sources as well. The extent to which sound in the sea impacts and affects marine life is a topic of considerable current interest both to the scientific community and to the general public. Scientific interest arises from a desire to understand more about the role of sound production and reception in the behavior, physiology, and ecology of marine organisms. Anthropogenic sound, including sound necessary to study the marine environment, can interfere with the natural use of sound by marine organisms. Public interest arises primarily from the potential effects of anthropogenic sound on marine mammals, given the broad recognition of the importance of sound in the lives of marine mammals.

For acoustical oceanographers, marine seismologists, and minerals explorers, sound is the most powerful remote-sensing tool available to determine the geological structure of the seabed and to discover oil and gas reserves deep below the seafloor. Society as a whole has reaped substantial intellectual and practical benefits from these activities, including bottom-mapping sonars and technology leading to the discovery of substantial offshore oil reserves.

Scientists and the public are also acutely aware that sound is a primary means by which many marine organisms learn about their environment and that sound is also the primary means of communicating, navigating, and foraging for many species of marine mammals and fish. Indeed, the study of sounds of marine organisms provides insight into important aspects of their biology.

The public's interest in the impact of human-generated ocean noise on marine animals has greatly increased. Concerns include whether human-generated sounds may interfere with the normal use of sound by the marine animals or whether the human-generated sounds may cause the animals physical harm. At issue is whether the human-generated sounds affect the ability of marine animals to pursue their normal activities and the long-term ability of these animals to survive, reproduce, and maintain healthy populations.

It is also critical to note that sound is an essential tool for ensuring national security. The development of underwater sound as a method for detecting submarines began during World War I and accelerated rapidly during World War II. During the Cold War, acoustic antisubmarine warfare became the principal deterrent against missile-carrying submarines roaming the high seas. Since the end of the Cold War ocean acoustics has continued to retain its military significance, but now militaries seek to expose submarine and submerged mine threats in shallow-water areas.

It is in this context of parallel developments and applications in ocean acoustics, marine seismology, oil exploration, and animal bioacoustics that concerns about the effects of sound on marine life have emerged. While researchers had been aware for quite some time of the sounds produced by marine life, it was not until the Acoustic Thermometry of Ocean Climate (ATOC) project (Baggeroer and Munk, 1992), in which high-intensity, low-frequency (defined for this report as sounds below 1,000 Hz) sounds were transmitted over long distances, that the public's attention focused on the possible impacts of human-generated noise on marine mammals, although noise with potential impacts had been regulated since the passage of the Marine Mammal Protection Act in 1972. Suddenly, it seemed, nearly all sources of anthropogenic sound came under intense scrutiny as potential threats to the existence and well-being of undersea life. These have included not only the aforementioned oceanographic, naval, and seismic surveying tools but also additional sources of unintentionally generated noise, such as commercial shipping, offshore construction, and recreational boating. As a result, research support for marine mammal bioacoustics, principally from the Office of Naval Research (ONR; Gisiner, 1998), grew substantially, and the permitting process necessary for conducting ocean acoustics experiments that allow incidental takes, administered by the National Oceanic and Atmospheric Administration (NOAA) and the U.S. Fish and Wildlife Service, received increased scrutiny. Two National Research Council (NRC) panels (NRC, 1994, 2000) were convened especially to address those issues associated with low-frequency sound, with particular attention paid to the ATOC project (NRC, 2000). The current NRC committee, which is responsible for generating this report, was convened at the request of the interagency National Ocean Partnership Program, with support from ONR, the National Science Foundation, NOAA, and the U.S.

Geological Survey. It was requested in the context of growing concern over noise in the ocean [Natural Resources Defense Council (NRDC), 1999] and with the recognition that there was a need to focus on a broader range of issues than those associated with the ATOC project.

Although the thrust of this study and those that have preceded it (NRC, 1994, 2000) is the impact of anthropogenic sounds, it must be realized that sound in the sea is produced by a large and extraordinarily diverse number of naturally occurring nonbiological and biological sources. Natural non-biological sounds are as diverse as the wind and waves, rockslides, geologic events, thunderstorms, and water moving over a coral reef. Many of these sources of sound have existed since the formation of the earth and oceans, and it is highly likely that these sounds have had some impact on the evolution of the auditory system, animal communication, and ecology (Fay and Popper, 2000). Biologic sounds are equally diverse and are emitted intentionally or unintentionally by numerous organisms. Unintentional sounds include, for example, those produced by schools of fish swimming through the ocean or release of air by large groups of fish as they adjust their buoyancy (Moulton, 1960, 1963). Intentional sounds, including whale songs, dolphin clicks, and fish vocalizations, are believed to be produced in various species for communication, echolocation, and perhaps even acoustic "imaging" of the environment to assess the physical characteristics of their habitat.

Sound detection by vertebrates clearly arose in the aquatic environment (Fay and Popper, 2000). The earliest known vertebrate fossils had ears (Jarvick, 1980), although there is no way of knowing if these ears functioned for sound detection or only served for detection of head motion and balance. Ears and functioning auditory systems are found in all aquatic vertebrates.[1] Auditory capabilities of bony fish are reasonably sophisticated, and a number of species not only detect sounds but can also determine sound source direction, detect signals in the presence of noise sources (maskers), and discriminate between sounds (e.g., Popper and Fay, 1999; Fay and Popper, 2000). Moreover, there is considerable similarity in the structure of the ear in aquatic and terrestrial vertebrates, and it is clear that the basic structure of the ear, including the sensory hair cell that converts sound to signals in the nervous system in all vertebrates, evolved very early in vertebrate history (see Popper and Fay, 1997; Fay and Popper, 2000).

The questions then to ask are *why* hearing evolved and *why* one would

---

[1]The only exception may be the jawless fish, lampreys and hagfish, where there is a functioning ear but no evidence to indicate whether they can or cannot detect sound. In these species, the ear may strictly serve as an organ of balance.

expect hearing to be particularly sophisticated in marine animals.[2] The aquatic environment has limited or no light, and even in areas where there is considerable light, the range of visibility is rather limited as a result of the attenuation characteristics of light in water. As a consequence, if early aquatic animals had only visual systems, the range of information about the environment around them would have been constrained by their field of vision. With the evolution of the auditory system, the sensory world of the organism expands to greater distances and the animal develops an acoustic image of the world around it, just as humans sense the world around them using sound, even when vision is not available. The evolution of an auditory system that can discriminate among sounds, determine the direction of a sound source, and detect sounds even when the environment is reasonably noisy greatly increased the survival potential of aquatic animals. It has been argued that humans and animals glean a great deal about their environment from the "acoustic scene" and that this scene provides an immense amount of subtle information (see Bregman, 1990; Fay and Popper, 2000). Indeed, Bregman's ideas can be extended to argue that the most important aspect of hearing is not communication per se but learning about the acoustic scene in order to detect objects and organisms in the environment and the ability to discriminate between sounds and the location of different sounds, a process called "stream segregation" (Bregman, 1990; Fay and Popper, 2000).

In essence, sound and sound detection would seem to be critical parts of the lives of marine mammals and fish.[3] Many of these animals use sound for communication between members of their species. But equally important is the idea that probably *all* of these species use sound to learn about their environment and to survive. Therefore, there should be concern not only about the impact of anthropogenic sounds on communication but also about the impact on general determination of information in the environment.

A fundamental question is whether the impact of anthropogenic sounds on marine mammals and the marine ecosystem is sufficiently great to warrant concern by both the scientific community and the public. As discussed in detail in this report, the data currently available suggest that such interest is indeed justified. However, as will also be shown, the data are still quite limited, and it will be important to develop a research program that will

---

[2]*How* the ear evolved is another issue of considerable interest, but one that will not be considered here. Readers are referred to van Bergeijk (1967), Baird (1974), Ridgway et al. (1974), and Fay and Popper (2000) for useful discussions of this issue.

[3]It should be noted that there have been very few studies on sound detection by marine invertebrates and so we do not yet know if any of these species detect sound.

provide substantially more data on this topic. Only when these data are available will it be possible to draw concrete conclusions regarding this question. The statement of task and the committee's response provide the framework for obtaining these data.

STATEMENT OF TASK

This study will evaluate the human and natural contributions to marine ambient noise and describe the long-term trends in ambient noise levels, especially from human activities. The report will outline the research needed to evaluate the impacts of ambient noise from various sources (natural, commercial, naval, and acoustic-based ocean research) on marine mammal species, especially in biologically sensitive areas. The study will review and identify gaps in existing marine noise databases and recommend research needed to develop a model of ocean noise that incorporates temporal, spatial, and frequency-dependent variables.

In its interpretation of the statement of task, the committee felt that there were several key guidelines that should be followed and several key questions that must be addressed. First, to researchers the term "ambient noise" typically refers to the overall background noise caused by all sources such that the contribution from a specific source is not identifiable. For example, considering only shipping noise in this context, Cato (2001) states that "traffic [shipping] noise is the low-frequency general background noise resulting from contributions from many ships over an ocean basin, but in which the contribution of no individual ship is distinguishable." However, the committee felt that this conventional definition was too restrictive and that sound caused by identifiable, often transient, typically nearby sources should be included in its considerations as well. The term "ocean noise" was therefore defined by the committee as encompassing not only the usual background ambient noise but also the noise from distinguishable sources (Box 1-1). Second, the committee agreed that, although its work would concentrate primarily on the effects of noise on marine mammals, it should consider other species as well (e.g., fish) that are part of the ecosystem and food web on which marine mammals depend. Third, the frequency band to be studied was determined to range from 1 to 200,000 Hz (200 kHz), since this is the entire bandwidth that various marine organisms are capable of detecting.

Five key questions were considered to be essential to achieving the goals described in the statement of task:

1. What is the noise budget in the ocean?

It is well known that noise in the ocean arises from a variety of sources, including ships, breaking waves, and living organisms. Far less is known about the relative contributions of each of these sources (referred to, in this

---

Box 1-1
Sources of man-made noise in the ocean

TRANSPORTATION
Aircraft (fixed-wing and helicopters)
Vessels (ships and boats)
Icebreakers
Hovercraft and vehicles on ice

DREDGING AND CONSTRUCTION
Dredging
Tunnel boring
Other construction operations

OIL DRILLING AND PRODUCTION
Drilling from islands and caissons
Drilling from bottom-founded platforms
Drilling from vessels
Offshore oil and gas production

GEOPHYSICAL SURVEYS
Air-guns
Sleeve exploders and gas guns
Vibroseis
Other techniques

SONARS
Commercial sonars (including fish finders, depth sounders)
Military sonars

EXPLOSIONS

OCEAN SCIENCE STUDIES
Seismology
Acoustic propagation
Acoustic tomography
Acoustic thermometry

SOURCE: Richardson et al., 1995. Courtesy of Academic Press.

---

report, as the noise budget) to the total noise field in various parts of the world's oceans, including seasonal differences, or about the more detailed spatial and temporal variability of the noise field. Furthermore, within a particular source category (e.g., ships, seismic surveys) the contribution from subsets should be understood. For example, within the major category of ships the contribution from different types of vessels has not been quantified.

2. What are the long-term trends in noise levels?

It is clear that prior to the Industrial Revolution (ca. 1850), the contribution of anthropogenic activity to the noise budget was negligible and that ocean noise levels were determined by naturally occurring sources (e.g., wind, waves, earthquakes, organisms). Little is known about the changes of these levels with time as a result of the increased maritime activity associated with the onset of industrialization. To what extent has this trend been influenced by factors such as the number of ships, their size, and propulsion? In more recent years, changes in the noise budget would also have to take into consideration other sources of anthropogenic sounds discussed in this report.

In order to understand long-term changes in the noise budget caused by human activity, a baseline can be obtained from noise measurements in areas with few human-generated contributions, for example, several places in the southern hemisphere far removed from shipping lanes and where low-frequency sound from long range is blocked by bathymetry.

3. Are existing models of ocean noise still valid?

Probably the most widely used models of the ambient component of ocean noise continue to be the curves developed by Wenz (1962; see also Richardson et al., 1995). These provide a summary of average ambient noise spectra from various sources, as shown in Plate 1. But according to Ross (1993), "they are not particularly useful in predicting or explaining ambient noise measured in a particular location at a particular time." Furthermore, considerable additional noise data have been acquired and theoretical developments have occurred during the past 40 years (Gisiner, 1998), so that updated and improved versions of the Wenz curves could be developed. What are the effects of specific properties of noise sources, including rise times, tonal content, bandwidth, and power levels?

4. What are the effects of transient and long-term noise exposure on marine mammals and the ecosystems on which they depend?

Specific conclusions on the effects of noise-induced hearing loss on terrestrial mammals have been drawn. Recent experiments have shown that (1) the noise need not be painful to cause permanent loss; (2) the damage is approximately proportional to noise energy integrated over time; (3) high-frequency noise is more dangerous than low-frequency noise; (4) narrowband noise is more dangerous than broadband noise; and (5) there is large intersubject variability in the resistance to noise, even among genetically identical animals (Liberman, 2001). Comparable data are not available for marine mammals, although it is clear that such data are needed in order to understand the impact of anthropogenic sound on these organisms.

Despite the lack of data for marine mammals, some general comments can be made about the impact of noise on aquatic organisms by introducing the concept of zone of influence (Richardson et al., 1995; Gisiner, 1998;

NRDC, 1999). Essentially, the effect of noise on the animal depends to a large degree on the proximity of the animal to the noise source and the received level of the signal by the animal. At very short ranges (that have yet to be determined), a sufficiently loud source may cause severe physiological damage and perhaps death. At greater ranges, geometrical spreading and absorption reduce the signal level substantially and the same source may cause hearing loss and short-term behavioral changes, which can contribute to death under particular circumstances (Evans and England, 2001). A quantitative evaluation of the radii of these zones for different species as well as an understanding of effects analogous to those described for terrestrial mammals have yet to be determined.

It should also be noted that marine mammals are part of a larger ecosystem upon which they depend. Included in this ecosystem are other organisms, particularly fish and possibly marine reptiles and invertebrates, which use sound in their normal behavior and that may also be impacted by anthropogenic sounds. Thus, in addition to understanding the direct impact of such sounds on marine mammals, it is important to understand the impact of these sounds on fish and other organisms.

5. What are recommendations for future research?

None of the four preceding questions currently has a concrete and final answer. It is therefore crucial that specific areas for future research, leading to more conclusive answers, be identified. Research recommendations from previous NRC studies (1994, 2001) are included in Appendix D and should be reviewed and considered with those presented here. Progress has been made in many of the areas described in the previous reports, but much more must be accomplished to improve our ability to predict and assess the impact of ocean noise on marine mammals.

APPLICATIONS OF THE SONAR EQUATION TO BIOLOGICAL RECEIVERS

The following section presents the sonar equation and discusses its application to biological receivers. This section is not intended to be a thorough review of this topic but, rather, to introduce many of the terms and ideas that will be addressed throughout the remainder of this report. Additional terms along with measures of the properties of acoustic sources and acoustic fields are discussed in the Glossary. For additional study of the fundamentals of ocean acoustics and biosonar, interested readers can refer to one of several textbooks on these topics (e.g., Busnel, 1963; Urick, 1975; Tolstoy and Clay, 1987; Brekhovskikh and Lysanov, 1991; Burdic, 1991; Au, 1993; Frisk, 1994; Jensen et al., 1994; Richardson et al., 1995; Medwin and Clay, 1998).

The quantitative description of the acoustic pressure wave to which an animal is exposed is obtained through the use of the sonar equation (Urick,

1975; Jensen et al., 1994). Specifically, the received acoustic level (RL) from a source with source level (SL) is given by

$$RL = SL - TL + AG, \qquad (1\text{-}1)$$

where TL is the transmission loss from source to receiver and AG is the processing gain associated with the animal's reception system. All of the components of the sonar equation are expressed in decibels (dB), which are proportional to the logarithms of the corresponding linear values. The decibel is used largely for convenience, since the individual components of the equation may span a broad dynamic range, and furthermore, the logarithmic operation expresses multiplicative processes in terms of seemingly simpler additive operations. In addition, a logarithmic scale is typically used for sound levels because human perception of loudness increases logarithmically. Specifically, the decibel is inherently a relative quantity, that is

$$RL(\text{dB}) = 20 \ \log_{10}\left[\frac{\text{measured pressure}}{\text{reference pressure}}\right]$$

$$= 10 \ \log_{10}\left[\frac{\text{measured pressure}}{\text{reference pressure}}\right]^2 \qquad (1\text{-}2)$$

where the reference pressure level used in underwater acoustics is 1 µPa (see Glossary for further explanation). The SL is defined as the pressure at a unit distance, typically 1 m, from the source, while the TL describes all of the geometrical spreading and attenuating effects of the medium associated with propagation, scattering, and absorption as the signal travels from a position 1 m from the source to the location of the animal. The AG represents the enhancement of the received signal that can occur through the application of signal-processing techniques and perhaps multiple sensors in the receiving system. Combining all of these terms, the ability of the animal to detect the signal can be interpreted in terms of the animal's hearing sensitivity, that is, the minimum detectable value of RL, which expresses its minimum threshold[4] hearing level as a function of frequency (Figure 1-1) (Wartzok and Ketten, 1999; Finneran et al., 2002). In the

---

[4] It should be noted that the concept of the threshold is a statistical one and represents the minimal detectable level for an organism in some percent of trials—often 50 or 75 percent of trials. The threshold for an individual animal may change by a few decibels, even within the course of a testing session, and the threshold at any given moment may depend on motivational level and distractions in the environment (Holt et al., 2002).

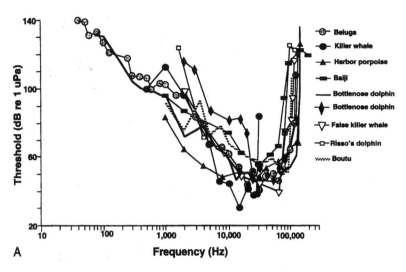

A

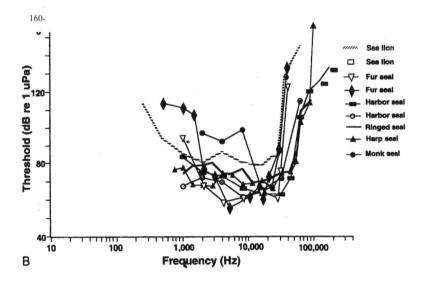

B

FIGURE 1-1 Audiograms for individual land mammals, cetaceans, and odontocetes. Underwater audiograms for (A) odontocetes and (B) pinnipeds. More than one curve is shown for some species because data reported in different studies were not consistent. Note that for both the bottlenose dolphin and the sea lion, thresholds are distinctly higher for one of the two animals tested. These differences may reflect different test conditions or a hearing deficit in one of the animals. SOURCES: Popper (1980), Fay (1988), Au (1993), and Richardson et al. (1995). Reproduced with permission from Wartzok and Ketten (1999). Copyright Smithsonian Institution Press.

ocean environment, an additional term must be introduced into the sonar equation, namely an ocean noise term (NL), which is defined with respect to the same reference pressure and frequency bandwidth as SL and RL. The actual excess signal level (SE) available to allow detection and interpretation of the signal is given by

$$SE = RL - NL = SL - TL + AG - NL. \qquad (1\text{-}3)$$

The animal will be able to hear and respond to a signal of a particular frequency only if SE is greater than zero. An interesting observation is that the superposition of odontocete hearing sensitivity (the audiogram) on the Wenz curves (Plate 2) indicates that the hearing thresholds of these animals correspond to the quiet ocean ambient noise spectral levels over the animals' frequency bands of hearing sensitivity. In other words, in the absence of human noise, the ocean is very quiet for them; they seem to have adapted to the natural noise that surrounds them.

Transmission loss in Equations 1-1 and 1-3 is a complicated function of the source and receiver geometry, frequency, and environmental parameters of the water column and the seabed (Brekhovskikh and Lysanov, 1991; Frisk, 1994; Jensen et al., 1994). In general, transmission loss with increasing source-receiver range is dominated by two important effects. First, the sound speed in the sea is not constant but varies with both depth and range, immediately altering the simple spherical spreading loss associated with a point source in free space. Sound waves interact with both the moving sea surface and the seabed, which is a complicated multilayered structure that supports acoustic waves. All of these factors combine to create a channel, or waveguide, for the sound waves that are trapped between the surface and the bottom in shallow water or focused by the sound speed structure in deep water as they propagate outward from source to receiver. This channeling effect causes the envelope of the signal to spread cylindrically, rather than spherically, outward at ranges much greater than the waveguide thickness, D (which equals the water depth in shallow water environments). Second, the intrinsic absorption properties of seawater cause the sound wave to be further attenuated by heat, viscous, and molecular relaxation losses (Medwin and Clay, 1998). As a result, the transmission loss can be expressed generally as:

$$TL \text{ (dB re 1 m)} = 20\log_{10} r + \alpha r, \text{ when } r < D \qquad (1\text{-}4)$$
$$TL \text{ (dB re 1 m)} = 10\log_{10} r + 10\log_{10} D + \alpha r - 3, \text{ when } r > D, \qquad (1\text{-}5)$$

where r is the horizontal range between source and receiver (in m), and the absorption coefficient α (in dB/m) is approximately proportional to the square of the frequency (Figure 1-2; Frisk, 1994) with the impact of absorption shown for an idealized case. Equations 1-4 and 1-5 are valid only for

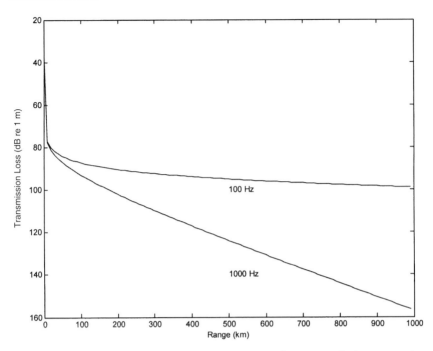

FIGURE 1-2 Ideal transmission loss. Transmission loss in an ideal 5,000-m-deep ocean with perfectly reflecting surface and bottom. This chapter details the calculation for transmission loss. The differences between the curves for 100 Hz and 1,000 Hz are due to frequency-dependent absorption by seawater.

omnidirectional, single-point sources; the geometrical spreading for other types of sources (e.g., line sources such as vertical source arrays) may be significantly different.

Waveguide effects are important in determining the distance traveled and the character of acoustic energy as it propagates through the ocean. The key factor that influences the character of the propagation in deep water is the variation with depth z of the sound velocity profile $c(z)$. Amazingly, the small relative variations in sound speed, which are typically less than 4 percent, have a profound influence on the structure of the sound field. Ducting by the sound speed structure dominates over any interactions with the boundaries in sound propagating from a deep source (about 1,000 m) in the classical SOFAR (sound fixing and ranging) channel found, for example, in the North Atlantic Ocean. The complexities of sound propagation in the sea must be carefully and accurately taken into account when evaluating the contribution of a particular sound source to the overall ocean noise field and are presented in more detail in Chapter 4.

In coastal regions and coral reefs where water depth is very shallow compared to that of the deep ocean, propagation of sound is more complex (Frisk, 1994). In these areas sound propagates over distances greater than a few water depths only by repeatedly interacting with the surface and bottom. At both the surface and bottom, a sound wave reflects back onto itself, and these reflections interfere with the original wave to produce an interference pattern in the water column. A sound source transmitting at a single frequency will produce a discrete number of vertical interference patterns, each with a different number of maximum and minimum pressures from top to bottom (Ferris, 1972). Each vertical interference pattern, or standing wave in the vertical direction, propagates in the horizontal direction at its own speed. However, if the frequency of a standing wave is too low, it will not propagate. This lower frequency limit is called the cutoff frequency, and standing waves with frequencies below the cutoff cannot propagate in the horizontal direction. Therefore, at a given water depth, an absolute cutoff frequency exists that is equal to the cutoff frequency for the vertical interference pattern having the fewest number of maximum and minimum pressures in the vertical (Rogers and Cox, 1988). A simple mathematical model of the shallow water environment can be devised by assuming it consists of a homogeneous ocean overlying a fluid-like, homogeneous bottom. For this model the absolute cutoff frequency (in Hz) below which no sound can propagate in shallow water, is given by

$$f_{cutoff} = \frac{c_w}{4h\sqrt{1 - \frac{c_w^2}{c_s^2}}} \qquad (1\text{-}6)$$

where c_w is the speed of sound in water, c_s is the speed of sound in the bottom sediment, and h is the water depth in meters. Real ocean bottoms are much more complicated than the simple homogeneous model described, and the bottom can become part of the medium in which the sound propagates (Figure 1-3). The propagation efficiency of the seabed, however, is far less than that of the water column because the intrinsic absorption of the bottom is typically about 1,000 times that in seawater. Because of variations in water depth and in ocean bottom properties (as well as variations in the sources of noise themselves), ocean noise in shallow water can be highly variable from one location to another (Urick, 1984; Zakarauskas, 1986).

In many cases of waveguide propagation in the ocean, the upper boundary of the waveguide is formed by reflection from the underside of the ocean surface. Therefore, the sea surface plays a fundamental role in acoustic propagation. Interaction of sound with the ocean surface also is important from a biological perspective, since marine mammals must come

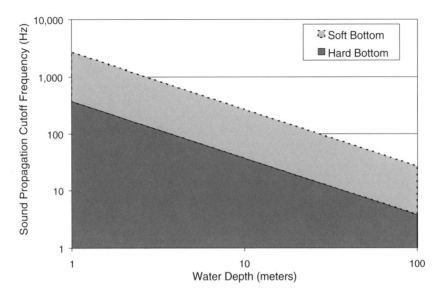

FIGURE 1-3   Cutoff frequencies estimated for propagation of sound in shallow water environments composed of a homogeneous ocean overlying a fluid-like, homogenous bottom. Sound at frequencies below the cutoff frequency (indicated by the shaded regions) will not propagate in the horizontal direction. The speed of sound in water is assumed to be 1,500 m/s. Speed of sound in the soft bottom is 1,520 m/s and 5,000 m/s in the hard bottom. Cutoff frequency was calculated using Equation 1-6.

to the surface to breathe. The sea surface under calm conditions is a nearly perfect reflector of ocean-borne sound at all incident angles over a wide band of frequencies.[5] Because the overlying mass of air provides very little resistance to particle motion (its acoustic "impedance" is small compared to that of seawater), the sea surface yields completely to the incoming underwater sound field. At this interface the ocean acoustic particle motion in the vertical direction is maximum and the acoustic pressure becomes zero, known as pressure release. Actual open-ocean surface conditions are complicated by factors such as the presence of near-surface bubbles and moving, wind-generated roughness. Animals that sense acoustic pressure can reduce their received sound levels by going to the ocean surface. As a result, comparisons of the density of marine mammals near sound sources and in other locations where the underwater sound levels are high may be

[5]Although underwater sound incident on the underside of a flat ocean surface is perfectly reflected for all intents and purposes, airborne sound that is nearly vertically incident on the sea surface can couple into ocean-borne sound, as discussed in Chapter 2.

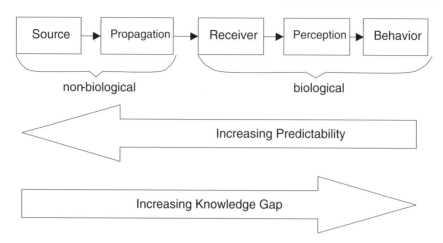

FIGURE 1-4 Components necessary to understand the effects of ocean noise on marine mammal behavior.

biased by the animals moving close to the surface in the presence of the sound in order to reduce the received sound pressure level.

STRUCTURE OF THE REPORT

This report describes sound origins, trends, effects on marine mammals, and current modeling efforts (Figure 1-4). Chapter 2 provides descriptions of the natural and human sources of ambient noise in the ocean and the possible reasons and evidence for long-term trends in ocean noise. Chapter 3 describes what is known about the impacts of marine noise on mammals, including masking, sensitization, and habituation. Chapter 4 summarizes existing modeling efforts and ocean noise databases, particularly those that integrate the known information about noise with behavioral databases on marine life. Chapter 5 synthesizes findings and recommendations of the committee for future research.

2

Sources of Sound in the Ocean and Long-Term Trends in Ocean Noise

INTRODUCTION

In this chapter the major natural (physical and biological) and anthropogenic contributors to ocean noise are discussed. Gaps in our knowledge or available data are identified that will need to be addressed in future research in order to develop predictive models of the effects of noise on marine mammals. A more thorough description of modeling efforts is contained in Chapter 4.

This chapter focuses on the properties of the sources and does not describe in detail the effects on the environment as the acoustic energy travels away from the vicinity of the sources. Parameters such as source level (in units of dB re 1 μPa at 1 m), source spectral density level (units of dB re 1 μPa^2 per Hz at 1 m), and time-integrated source pressure amplitude squared for use with transient signals (units of dB re 1 μPa^2 at 1 m) are presented for many of these sources, particularly man-made sources. However, accurate estimation of the source properties for many types of naturally occurring sounds is impossible, given the lack of knowledge of the individual source locations, of the spatial distribution of multiple contributing sources, and of the complex propagation conditions. Therefore, in such situations, the measured properties of the received acoustic field (which are obtained directly and require no additional information, computation, or assumptions, but which contain the effects of propagation) will be presented. The text clearly differentiates between the properties of the sources and those of the received field. The distinction between source level and received level also is discussed both in Chapter 1 and in the Glossary.

In the absence of shipping, natural forces are the dominant sources of the long-term time-averaged ocean noise at all frequencies. In the presence of distant shipping, contributions from natural sources continue to dominate time-averaged ocean noise spectra below 5 Hz and from a few hundred hertz to 200 kHz. The dominant source of naturally occurring noise across the frequencies from 1 Hz to 100 kHz is associated with ocean surface waves generated by the wind acting on the sea surface. Nonlinear interactions between ocean surface waves called microseisms (see the Glossary; referred to as "Surface Waves—Second-Order Pressure Effects" in Plates 1 and 2) are the dominant contributors below 5 Hz, while thermal noise (i.e., the pressure fluctuations associated with the thermal agitation of the ocean medium itself) is the dominant contributor above 100 kHz. Natural biological sound sources make a noticeable contribution at certain times of year. For example, a peak around 20 Hz created by calls of large baleen whales is often present in deep-ocean noise spectra. Groups of whistling and echolocating dolphins can raise the local noise level at the frequencies of their signals. Snapping shrimp are an important component of natural noise from a few kilohertz to above 100 kHz close to reefs and in rocky bottom regions in warm shallow waters. Fish can add to ocean noise in some locales.

Whether intentional or unintentional, anthropogenic noise in the marine environment is an important component of ocean noise. Sound is a widely used tool for a broad range of marine activities. In the search for new hydrocarbon reserves, the rock underlying the seafloor is characterized using air-guns. Marine researchers use sound waves to investigate the properties of seawater both for local and global studies. Sonars used for civilian navigation and defense purposes use sound waves to locate objects under the sea surface. Unintentional contributions to marine noise arise from transiting ships, coastal and marine construction activity, mineral extraction, and aircraft overflights. These anthropogenic sound sources contribute to ocean noise over the complete 1-Hz to 200-kHz band of interest in this report. In the lowest bands, 1-10 Hz, the contributors are ship propellers, explosives, seismic sources, and aircraft sonic booms. In the 10-100 Hz band, shipping, explosives, seismic surveying sources, aircraft sonic booms, construction and industrial activities, and naval surveillance sonars are the major contributors. For the 100-1,000 Hz band, all the sources noted for the 10-100 Hz band still contribute. Also, the noise from nearby ships and seismic air-guns can extend up into the 1,000-10,000 Hz band. This band also includes underwater communication, naval tactical sonars, seafloor profilers, and depth sounders. The 10,000-100,000 Hz band includes the systems listed, in addition to mine-hunting sonars, fish finders, and some oceanographic systems (e.g., acoustic Doppler current profilers). Anthropogenic contributors at and above 100,000 Hz are limited to mine hunting, fish finders, high-resolution seafloor mapping devices such as side-scan sonars, some depth sounders, some oceanographic sonars, and research sonars for small-scale oceanic features (Table 2-1a and 2-1b).

Prior to considering anthropogenic sources, it is useful to first understand the natural sources that contribute to ocean noise. Presumably, hearing and communication systems of marine organisms are adapted to these natural noises.

NATURAL SOURCES OF OCEAN NOISE

Physical and Geophysical Sources

The ocean is intimately coupled to the solid earth and the atmosphere, and in fact, most of the significant physical sources of natural sound occur at the interfaces among these three media. Additional sound in the marine environment originates in the atmosphere and penetrates the ocean surface. Elastic vibrations in the earth also introduce sound into the underwater acoustic field.

Sources at the Ocean Surface

The dominant physical mechanisms of naturally occurring sound in the ocean occur at or near the ocean surface. Most are associated with wind fields acting on the surface and the resulting surface wave activity. In the absence of man-made, biological, and transient sounds, ambient noise is wind dependent over the band from below 1 Hz to at least 50 kHz. Below 5-10 Hz, the dominant ambient noise source is the nonlinear interaction of oppositely propagating ocean surface waves. These sounds are called microseisms. (The term "microseisms" comes from the fact that they also are the dominant source of noise in high-quality, on-land seismometer measurements; however, the source mechanism for microseisms is unrelated to seismic processes in the solid earth.) Across most of the remainder of this band, the primary sources are bubbles that are oscillating, both individually and collectively in a cloud, in the water column. Several good references on natural physical sources of ocean noise and the properties of the ambient noise field are available (e.g., Urick, 1984; Zakarauskas, 1986; Ross, 1976; Kerman, 1988, 1993; Buckingham and Potter, 1995; Leighton, 1997; Deane, 1999). Only a brief summary of the major contributors to the underwater sound field is given here. However, in some frequency bands such as the band from 10 to 200 Hz, where ambient noise in the northern hemisphere typically is dominated by shipping noise, the dominant source mechanisms have not been identified. Quantification of the relative contributions of the various mechanisms of naturally occurring sound created at the sea surface remains an active area of research.

The average ocean noise spectrum can be empirically described and

TABLE 2-1a Characteristics of Anthropogenic Contributors to Marine Noise

SOURCE	SPATIAL VARIABILITY			DIRECTIONALITY	ACTIVITY LEVELS	
	Large Scale Ocean Basin to Global	Mid Scale 10s of km to Ocean Basin	Small Scale <1 km to 10s of km		Number of Regional Sources	Frequency of Activity in Region
Shipping	**Presence is global for all types and limited to the ocean surface**					
Merchant	Shipping lanes transcend ocean basins and are populated continuously				1-2	4-5/hr
Utility		Operations are confined to subocean basins and localized areas such as fishing grounds		All shipping: Generally considerd to be omnidirectional, but shielding is certainly present in the horizontal plane, especially for the higher frequencies; omnidirectional in the vertical plane.	1-30	daily
Military		Operations are military exercises, war zones	Extending down to amphibious assault zones, beach heads		6-10	bi-monthly

				Directionality	Number	Rate
Scientific		Specific sites to observe phenomena of limited spatial scales	Down to localized phenomena such as "black smokers"		1-2	monthly
Recreation			Coastal regions, limited range		>10	>5/day
Other	E.g., transoceanic cable laying	Localized operations	E.g., drill site			>monthly
Seismic exploration		Surveys to >100 km	Down to 10s of km	Omnidirectional	1	monthly
Sonars	**Global presence, but variability is defined by sonar use**					
Military Surveillance		Ocean basin use	Down to 10s of km	Omnidirectional	1	monthly
Tactical		10s of km and up, conditions permitting	Down to 10s of km	Horizontal plane, Vertical >100°, Vertical plane <20°	2-3	See host platform data above
Weapon/ Counter Weapon			10 m to >10 km	Highly directional in both planes (<5°)	1-2	
Civilian Communications		10s of km and up	Down to >1 km	Horizontal plane: omni Vertical plane: <10°	1	

Continued

TABLE 2-1a *Continued*

SOURCE	SPATIAL VARIABILITY Large Scale Ocean Basin to Global	Mid Scale 10s of km to Ocean Basin	Small Scale <1 km to 10s of km	DIRECTIONALITY	ACTIVITY LEVELS Number of Regional Sources	Frequency of Activity in Region
Navigation			<1 km to <10 km	Omnidirectional	1	
Hi-resolution			>10 m to >100 m	Highly directional (almost all look down or up)	1-2	
Marine Research	Limited ocean basin tests	Spatial interest to ocean basin dimensions	Down to sub-meter measurements	*See sonars / military and civilian*	1	monthly
Explosions			<1 km	*Omnidirectional*	1	seldom
Industrial Activity	Presence is global and limited to near-shore locations/onshore locations					
Construction			<1 km to 10s of km	Omnidirectional	1	
Dredging			<1 km to 10s of km	Omnidirectional	1	
Power plants			100s of m	Omnidirectional	1	
Factories			100s of m	Omnidirectional	1	

Transportation	10s of km upward	<1 km to 10s of km	Road noise – unknown, Ferries – same as ships	1–2	Up to 1/hr
Miscellaneous **Global presence**					
Aircraft overflight	Very similar in extent and character to shipping lanes		*See text-confined to vertical cone*	1	4-5 hrs
Military activity nonsonar		< 1 km	*Omnidirectional*	1	seldom

Note: Spatial variability is an indicator of the geographic distribution of the sources. Directionality refers to the direction in which the signal is projected. Activity levels indicate the likely number of regional sources, and the frequency of signal occurrence within that region.

TABLE 2-1b Characteristics of Anthropogenic Contributors to Marine Noise

SOURCE	TEMPORAL VARIABILITY			SOURCE CHARACTERISTICS		
	Large Scale wks to mos	Mid Scale hrs to days	Small Scale sec(s) to min(s)	Signal Structure	Spectral Content	Source level dB re 1 μPa at 1 m
Shipping				See Figure 2-1 for example		
Merchant	Constant presence			occasional transient due to operations activity on vessel		160–220
Utility	On site for wks	Down to hrs		numerous transients due to nature of operations	For all shipping, broadband energy from 10 Hz to >1 kHz with spectral lines rising above B/B due to propulsion blades, turbines, generators, etc. Transient additions are broadband and flat from 10 Hz to 1 kHz	160–200

160–200 |

				Frequency	Source level
Military	On site for hrs to days		general level up and down with exercise/war fighting requirements		160–220
Scientific	On site for days	Down to min(s)	stop and start behavior driven by data collection schedule		160–200
Recreation Other	Wks Typically hrs days /hrs	Or less	highly variable		160–190
Seismic exploration	On site for days		Impulsive (see Figure 2-4)	Broadband	>240
Sonars *Military*					
Surveillance	On site for wks	Down to days	Pulsed tones	<1 kHz	>230
Tactical	On site for hrs	Down to min(s)	Pulsed tones	>1 kHz to <10 kHz	200 to 230
Weapon/ Counter Weapon	hrs to ~ a day	Down to min(s)	Pulsed tones / Wideband pulses	>0 kHz to >100 kHz	190 to 220

Continued

TABLE 2-1b *Continued*

SOURCE	TEMPORAL VARIABILITY			SOURCE CHARACTERISTICS		
	Large Scale wks to mos	Mid Scale hrs to days	Small Scale sec(s) to min(s)	Signal Structure	Spectral Content	Source Level dB re 1 µPa at 1 m
Civilian						
Communications		hrs	To min(s) tones	CW/Pulsed	Low kHz to >10 kHz	180-210
Navigation		hrs/days	min(s)	CW/Pulsed	Low kHz to >10 kHz	180-210
Hi-resolution			min(s)	Pulsed tones	>10 kHz to >100 kHz	160-220
Acoustic				Series of pulses 10-500 msec w/ interpulse periods of 0-10 sec	5-30 kHz typically	130-150
Harassment or Deterrent Devices (AHD, ADD)						
Marine Research	Down to days/hrs			See military/civilian sonars	See military/ civilian sonars	160-220
Explosions			sec(s)	Impulsive (see figure)	Broadband	>240

Industrial Activity

Construction	wks to mos	Down to days	Broadband and Tones/CW	<10 Hz to 1 kHz	Unknown
Dredging	wks to mos	Down to days	Broadband and Tones	<10 Hz to <1 kHz	Unknown
Power plants	Constant process		CW and some transients	<100 to several 100 Hz	Unknown
Factories	Constant process		CW and significant transients	<100 to several 100 Hz	Unknown
Transportation	Constant process		Same as shipping and highway noise	See shipping and broadband	170–210

Miscellaneous

Aircraft overflight	Constant process		Both CW and broadband	<100 Hz to 10 kHz	Unknown
Military activity nonsonar	hrs	*Down to sec(s)*	Impulsive and broadband	Broadband	Unknown

Note: Frequently reported source signal characteristics are given although additional source characteristics, such as rise time, may also be important in determining the effects of the sources on marine mammals. This table is not meant to be a catalog nor does it approach all-inclusiveness but is provided to give a sense of the breadth of human impact on the undersea environment.

parameterized according to sea state (Knudsen et al., 1948). These Knudsen curves are straight lines of spectral density as a function of frequency when plotted on a logarithmic scale. The parallel nature of the "curves" for various sea states signifies that the noise level increases with increasing sea state by the same amount at all frequencies. Although developed more than a half-century ago, the Knudsen curves continue to be widely used to predict natural ocean noise levels at frequencies from 1 to 100 kHz. The pioneering Knudsen's curves of noise as a function of sea state have been very useful for many years and are remarkably effective, but it is now well established that the noise is correlated much better with wind speed than with sea state or wave height (correlation of wind speed and sea state only occurs in equilibrium conditions). This correlation with wind speed allows much more effective prediction and forecast (from wind forecasts) than could be obtained from sea state, which is difficult to estimate reliably.

Although open-ocean breaking wave noise is correlated with wind speed, local winds are not required to create the sounds from breaking surf. The sound created by spilling breakers (breaking begins at the wave crest and proceeds down the face of the wave) is primarily at the higher frequencies, whereas that from plunging breakers (the water at the wave crest leaps ahead of the wave in a jet, encompassing a large column of air) is significantly greater in levels and in frequency bandwidth. Plunging surf can raise underwater noise levels by more than 20 dB a few hundred meters outside the surf zone across the band from 10 Hz to 10 kHz (Wilson et al., 1985).

Precipitation on the ocean surface also contributes sound to the ocean. Rain can increase the naturally occurring ambient noise levels by up to 35 dB across a broad range of frequencies extending from several hundred hertz to greater than 20 kHz. For drizzle in light winds, a broad spectral peak 10-20 dB above the background occurs near 15 kHz (Nystuen and Farmer, 1987; measurements made at 7.5 m depth in an 8 m deep spot in a soft-bottom lake, Nystuen, 1986).

Atmospheric Sources

Sounds originating in the atmosphere can couple into the underwater sound field. However, because of the large difference between the speed of sound in air and in water, the received underwater acoustic levels are highly dependent on the position of the underwater receiver relative to the atmospheric source. That is, for a range-independent ocean with a smooth ocean-air interface, only atmospheric sources within a 13° cone about the vertical above the underwater receiver are well coupled into underwater sound fields that can propagate to the receiver. Actual environmental and propagation conditions can complicate this simple picture and may allow sound originating outside the 13° cone to be audible (see Sparrow, 2002, for comments on the relative importance of some of these effects). The properties of atmospheric sound sources and characteristics of propagation limit their important contributions to the underwater sound field to low and infrasonic frequencies.

Thunder and lightning are one example of a naturally occurring atmospheric source of ocean noise. Underwater recordings of spectra of a received sound of thunder from a storm 5-10 km away show a peak between 50 and 250 Hz up to 15 dB above background levels, with detectable energy down to 10 Hz and up to 1 kHz (Dubrovsky and Kosterin, 1993). When surface ducting conditions exist (i.e., the sound speed increases with depth from the surface), this low-frequency energy can couple into the duct and propagate for very long distances in the ocean.

Other naturally occurring sources are auroras and supersonic and exploding meteoroids (bolides). Rough estimates indicate that at least one bolide event with the equivalent explosive yield of 15 kilotons of TNT occurs in the earth's atmosphere per year (ReVelle, 2001).

Geologic Sources

Seismic energy created by earthquakes can couple into acoustic waves in the ocean and travel over great distances. All types of tectonic processes, including subduction, spreading, and transform faulting along the midocean ridges and associated earthquake, volcanic, and hydrothermal vent activity, are found below the oceans and along their margins. These processes can make significant contributions to the marine noise field (Box 2-1). At short ranges, underwater sounds from earthquakes can extend to frequencies greater than 100 Hz. The arriving signal can have a very sharp onset, similar to that from an explosion, and can last from a few seconds to a few minutes. T phases, earthquake arrivals whose propagation pathway is predominantly through the ocean, recorded at long distance from the earthquake source region typically contain a broad peak in their pressure spectrum centered around 5 Hz.

Movement of sediment by current flow across the ocean bottom can be a significant source of ambient noise at frequencies from 1 kHz to greater than 200 kHz (Thorne, 1986).

Effects of Ice

An ice cover at the ocean surface radically alters the ocean noise field. The impact varies according to the type and degree of ice cover, whether it is shore-fast pack ice, moving pack ice and ice floes, or at the marginal ice zone (Milne, 1967). The effects of the ice cover also are determined by the mechanical properties of the ice itself, which are dependent on temperature. Shore-fast pack ice can result in a significant decrease in ambient noise levels, 10-20 dB, by isolating the water column from the direct effects of

Box 2-1
Underwater earthquakes—How loud are they?

The sizes of earthquakes are commonly characterized by magnitude (Richter, 1958). The two types most often used in modern-day earthquake bulletins are the body wave magnitude and the surface wave magnitude (Aki and Richards, 1980). Both involve measuring the ground displacement, A, in microns (equal to one-millionth of a meter) and the period, T (time interval between two peaks or two troughs in the time series), for a specified portion of the recorded signal, and calculating the ratio, A/T. The logarithm to the base 10 then is taken of this ratio, similar to the calculation performed to obtain decibel units in acoustics.

Using a simplified approach, earthquake body wave magnitude, m_b, can be converted into equivalent decibels of underwater acoustic pressure.

$$m_b = \log(A/T) + Q,$$

where A is the ground displacement amplitude in microns (10^{-6} m) of a given arrival and T is its corresponding period in seconds. The quantity Q corrects for the focal depth of the earthquake and its distance from the receiver. The value of Q also contains the definition of an event that has a magnitude of zero at a given reference distance; for 1-s period (i.e., 1-Hz frequency) waves, a zero-magnitude earthquake has a 1-micron amplitude at 100 km from the source. Assuming the ground displacement is measured in the vertical direction at the ocean-bottom interface, the vertical displacement of the ground equals that of the water column. Recognizing that the vertical particle velocity, v_z, for a single-frequency arrival is roughly

$$v_z = 2\pi(A/T),$$

and using the relationship of acoustic pressure to acoustic particle velocity for a vertically traveling plane wave, that is

$$p = \rho^* c^* v_z$$

where ρ is the density of the water and c is the speed of sound, then the received level (RL) of sound in the ocean 100 km from an event with body wave magnitude, m_b, is

$$RL \text{ (dB re 1 } \mu\text{Pa)} = 139.5 + 20^* m_b.$$

This equation illustrates the similarity between earthquake magnitude and the decibel scale in acoustics. The actual coupling of earthquake-generated seismic energy into the underwater sound field is too complicated and variable from one earthquake to the next for this equation to apply generally.

wind, although the sound of wind crossing the ice surface can be transferred to the water column. Decreasing air temperatures can cause thermal stresses and result in tensile cracking of rigid ice, and diurnal variability in air temperatures is sufficient to change received sound levels by 30 dB between 300 and 500 Hz (Urick, 1984). The underwater sound pulses that are emitted typically are a few milliseconds in duration and so have broad spectral content from 100 Hz to 1 kHz. Though sound is created within moving ice packs from the relative motion of adjacent ice blocks, much higher amplitude sounds are released by cold, rigid ice from mechanical-stress-induced cracking. This cracking, analogous to earthquakes, releases transient signals that are different in character from those resulting from thermal cracking, often lasting a hundred times as long or more. The basin-wide summation of the noise from these fracture mechanisms appears to be the main cause of the broadband peak centered at 10-20 Hz, with spectral density levels of about 90 dB re 1 $\mu Pa^2/Hz$ in under-ice ambient noise measurements (Dyer, 1987; Makris and Dyer, 1986). The mechanical stresses involved in glacier calving and ice ridging also create very high levels of underwater sound (e.g., a pressure spectral density level of 97 dB re 1 $\mu Pa^2/Hz$ from 10 to 100 Hz was measured at 30 m depth and 100 m from an active ice ridge) (Buck and Wilson, 1986).

Within the marginal ice zone, the underwater noise is determined primarily by ocean surface wave activity (Makris and Dyer, 1991). The interaction of ocean waves with the ice edge creates noise levels 4-12 dB greater at 30 m than those in the open ocean, depending on whether the ice edge is sharp and compact (12 dB) or diffuse (4 dB) (Diachok and Winokur, 1974).

Biological Sources of Underwater Sound

Biological contributions to the underwater sound field are discussed in this section. This discussion is presented not only to help satisfy the committee's task of evaluating "the human and natural contributions to marine ambient noise" but also to provide an idea of how these sounds are similar to, or different from, natural sounds from physical sources and noise from anthropogenic sources. Once a full characterization of vocalization behavior, character, and distribution in time and space is available, it will provide a baseline for future studies of potential changes that might be indicative of adverse behavioral impacts from human-related stresses on the marine environment, such as chemical pollution, unintended fishing impacts, and coastal development, as well as man-made noise.

Characteristics of Marine Mammal Sound Production

Marine mammal sound production has been reviewed in several places (Watkins and Wartzok, 1985; Richardson et al., 1995; Wartzok and Ketten, 1999) and these reviews will not be repeated extensively here. Although the sounds generated by many marine mammals do not originate in their vocal cords, the term "vocalization" will be used as a generic term to cover all sounds discussed in this report that are produced by marine mammals. Marine mammal vocalizations cover a very wide range of frequencies, from <10 Hz to >200 kHz (Plate 3). Odontocetes, the dolphins and toothed whales, produce broadband clicks that can be characterized by species.

Peak energy is at frequencies between 1 and 200 kHz. Burst pulse click trains also can have peak energy well above 100 kHz and the constant frequency (CF) or frequency-modulated (FM) whistles range from 1 to 25 kHz, with harmonics as high as 100 kHz (Lammers et al., 2003).

Vocalizations of baleen whales (Mysticetes) are significantly lower in frequency than are those of odontocetes; frequencies are rarely above 10 kHz. Although there is a wide range of descriptors assigned to mysticete vocalizations, they can be broadly categorized as low-frequency moans (0.4-40 s with fundamental frequency well below 200 Hz); simple calls (impulsive, narrowband, peak frequency less than 1 kHz); complex calls (broadband pulsatile AM or FM signals); and complex "songs," in some cases with regional and interannual variations in phrasing and spectra. Infrasonic signals in the 10-20-Hz range are well documented in at least two species, the blue whale, *Balaenoptera musculus* (Cummings and Thompson, 1971), and the fin whale, *B. physalus* (Watkins, 1981). Suggestions that these low-frequency signals are used for long-distance communication and topological imaging of their environment are intriguing but have not been definitively demonstrated (Payne and Webb, 1971; Ellison et al., 1987).

The ability to use self-generated sounds to glean information about objects in the environment (echolocation) has been demonstrated in 13 species of odontocetes (Richardson et al., 1995). No odontocete has been shown to be incapable of echolocation. As outlined in the following, strong correlations exist between habitat types, societal differences, and peak spectra (frequencies at which the strongest signals occur) (Gaskin, 1976; Wood and Evans, 1980; Ketten, 1984). Based on their ultrasonic (echolocation) signals, odontocetes fall into two broadly defined acoustic groups: Type I, with peak spectra above 100 kHz, and Type II, with peak spectra below 80 kHz (Ketten and Wartzok, 1990). These categorizations are first-order approximations based on the predominant peak spectra of wild animals in their normal habitat. Several Type II species produce signals with peak energy at higher frequencies, for example, *Tursiops truncatus*, when tested in a high-noise environment (Au, 1993).

Type I echolocators are inshore and riverine dolphins that operate in acoustically complex waters. Amazonia boutu (*Inia geoffrensis*) routinely hunt small fish amid the roots and stems choking silted "varzea" lakes created by seasonal flooding. These animals produce signals up to 200 kHz (Norris et al., 1972). Harbor porpoises (*Phocoena phocoena*), a typical inshore species, use 110- to 140-kHz signals (Kamminga, 1988). Tonal communication signals are rarely observed in most Type I species (Watkins and Wartzok, 1985).

Type II species are nearshore and offshore animals that inhabit low-object-density environments, travel in large pods, engage in conspecific communication, and use lower-frequency echolocation signals. In seven odontocete species, ranging from the riverine dolphins such as *Sotalia fluviatilis*, through coastal species such as the bottlenose dolphin (*Tursiops truncatus*), to offshore species such as the spotted dolphins (*Stenella frontalis*), there is a negative correlation between body size and the maximum frequency of the whistles (Wang Ding et al., 1995). Many of the odontocete whistles have been described as "signature" calls identifying individuals (Caldwell and Caldwell, 1965), whereas burst-pulse sounds in killer whales are group specific (Tyack, 2000) and click codas in sperm whales are shared among individuals (Moore et al., 1993).

Source levels for cetacean vocalizations have been reported as high as 228 dB re 1 µPa at 1 m for echolocation clicks of false killer whales (*Pseudorca crassidens*) (Thomas and Turl, 1990) and bottlenose dolphins echolocating in the presence of noise (Au, 1993). The highest-level vocalizations are mature male sperm whale clicks with calculated source levels of 232 dB re 1 µPa at 1 m (Møhl et al., 2000). It is not surprising that the highest source level vocalizations are echolocation clicks since the animal is acoustically imaging its environment using the return echoes from some objects with low target strength. The high-frequency signals, which provide good spatial resolution, are rapidly attenuated as a result of high absorption losses. The short duration of an echolocating click (50-200 µs) (Au, 1993) means that the energy content integrated over time of the clicks is low even though the source levels are high.

Odontocete whistles have much lower source levels than echolocation clicks, ranging from less than 110 dB re 1 µPa at 1 m for spinner dolphins (*Stenella longirostris*) (Watkins and Schevill, 1974), to 169 dB for bottlenose dolphins (Janik, 2000), to 180 dB re 1 µPa at 1 m for short-finned pilot whales (*Globicephala macrorhynchus*) (Fish and Turl, 1976). The detection range for most vocalizations is estimated to be on the order of hundreds of meters and usually less than 1 km. Sperm whale vocalizations may be detected at ranges greater than 10 km (Watkins, 1980), and the highest source level, bottlenose dolphin vocalizations, have been estimated to be detectable by other dolphins under ideal conditions (low-frequency whistles, shallow-water spreading, sea state of 0) at ranges over 20 km (Janik, 2000), whereas vocalizations of Peale's dolphin (*Lagenorhynchus australis*) (Schevill and Watkins, 1971) can be detected at ranges of only a few tens of meters.

Mysticete vocalizations have the potential to be detected over long distances. Blue whales (*Balaenoptera musculus*) and fin whales (*B. physalus*) produce low-frequency (10-25 Hz) moans with estimated source levels up to 190 dB re 1 µPa at 1 m (Cummings and Thompson, 1971; Thompson et al., 1979). Accurate source-level estimates are difficult to make because of uncertainties in localizing the calling animal and in taking account of propagation effects. Source-level estimates of the low-frequency component of blue whale calls recorded in one experiment show a 10 dB spread about values of 170 dB re 1 μPa at 1 m (Thode et al., 2000). Vocalizations below 1 kHz with estimated source levels above 180 dB re 1 μPa at 1 m have also been recorded from most of the other large mysticetes such as the bowhead whale, the southern right whale, the humpback whale, and the gray whale (Richardson et al., 1995). Fin and blue whale vocalizations have been detected from ranges estimated to be greater than a hundred kilometers (Cummings and Thompson, 1971) to a confirmed range of 600 km for a blue whale using a large-aperture, multielement array (Stafford et al., 1998). Responses of conspecifics to these vocalizations have been observed only occasionally at ranges as great as 20-25 km (Watkins, 1981).

Source levels have been estimated for vocalizations of only a few species of pinnipeds. The highest levels are the underwater trills of Weddell seals (*Leptonychotes weddelli*), which can reach 193 dB re 1 μPa at 1 m (Thomas and Kuechle, 1982) and are a constant feature during the breeding season near Weddell seal colonies. These calls are easily detected by a hydrophone up to 4 km away, and one seal at an "isolated" man-made hole 4 km from the colony detected the calls, estimated the distance to the vocalizing seals through apparent ranging behavior, and swam the 4 km under ice to the colony where it was relocated (Wartzok et al., 1992).

Marine Mammal Contributions to Ocean Noise

Along the U.S. West Coast, the Navy's sound surveillance system (SOSUS) has recorded blue whale choruses in September and October that have increased the ambient noise up to 20 dB (Cummings and Thompson, 1994). Other species such as fin, humpback, or sperm whales also have the potential to increase the ambient in regional areas by a similar amount. Curtis et al. (1999) found that a strong annual peak in the 15-22 Hz band, with signal levels up to 25 dB above the baseline ambient noise level, was one of the clearest features in data collected over two years from bottom-mounted receivers at 13 widely distributed locations in the North Pacific. Whale sounds were detected in 43 percent of 170-averaged spectra collected once every five minutes. Contributions of noise by marine mammals can be significant over short periods of time and space in the middle of large assemblages of vocally active animals. Levels of broadband clicks and FM whistles can be so high within an active school of oceanic dolphins that nothing else can be heard. Typically, such conditions last less than an hour at a stationary hydrophone. On the other hand, in limited geographic areas, such as the underwater canyons off Kaikoura, New Zealand, sperm whales are continuously audible and a dominant acoustic feature (Gordon et al., 1992). In most regions, however, the vocalizations of cetaceans above 25 Hz are more transient phenomena, which, averaged over hours or days, do not make major contributions to the ambient noise field.

During breeding season cetacean contributions to marine noise increase substantially. Choruses of singing humpback whales were dominant features in the noise field during the spring breeding season at a single location 0.8 km off the coast of Maui, Hawaii, in 13 m of water (Au et al., 2000a). Highest sound levels were recorded during early March, at frequencies between 100 and 150 Hz, 250 and 350 Hz, and 600 and 650 Hz, coinciding with the peak of the breeding season. Time-averaged peak levels recorded about 2.5 km offshore reached 125 dB re 1 μPa (Au and Green, 2000).

Diurnal variation in vocal output of marine animals is commonly observed. Oceanic dolphins are typically more vocally active at night than during the day (Gordon, 1987; Goold, 2000). Singing male humpbacks were also found to be more vocally active at night than during the day (Au et al., 2000a). However, no evidence of diurnal variation in the vocal behavior of sperm whales has been observed (Gordon, 1987).

In general, pinniped vocalizations show a peak in occurrence during the breeding season. The most distinctive phocid vocalization in the high arctic is that of the bearded seal (*Erignathus barbatus*), whose song can often be heard on hydrophones when no seals are visible on ice floes. Bearded seal vocalizations may be heard up to 45 km from the source (Stirling et al., 1983).

Vocally active group-breeding pinnipeds, such as the walrus (*Odobenus rosmarus*), Weddell, and harp seal (*Phoca groenlandica*), concentrate so many vocalizing animals in a relatively small area that they can add to the local ambient background, although actual values of such increases are rarely reported. Harp seal breeding herds have been detected at over 2 km (Terhune and Ronald, 1986). The social stimulation of vocalizations between the herd and animals approaching the main herd and between the approaching animals and ones farther away leads to a much larger area in which vocalizations of individual harp seals can be detected. Terhune and Ronald (1986) reported hearing some harp seal vocalizations continuously along radii up to 60 km from the herd. The level of vocalizations varies such that fewer vocalizations are recorded on a second hydrophone located only a few hundred meters farther from either harp or Weddell seal herds than are recorded on the closer hydrophone (Terhune et al., 2001).

Other marine mammals such as the eared seals, manatees, dugongs, and sea otters have relatively low-level underwater vocalizations and add little to the acoustic scene. Except for the vocalizations of baleen whales, which can be detected for hundreds of kilometers, the contributions of marine mammals to the ocean sound ambient are localized in space. There is diurnal and seasonal variability in the occurrence of vocalizations, although in some locations marine mammal sounds are consistent features of the ambient. For example, hydrophones north of Oahu, Hawaii, recorded at least one whale sound on 459 of 578 recording days (Thompson and Friedl, 1982).

Ocean Noise from Fish and Marine Invertebrates

Many species of fish produce sound and use it for communication, and many more species produce sounds incident to other behaviors such as feeding and swimming (Busnel, 1963; Zelick et al., 1999; Box 2-2). The sounds are used in a variety of behavioral contexts, including reproduction, territorial behavior, and aggressive behavior (reviewed by Zelick et al., 1999). Elasmobranchs (sharks and rays) are not known to produce sounds, although they do respond dramatically to the sounds of potential prey (e.g., Myrberg, 1972; Myrberg et al., 1976) and are known to locate objects using sounds from over 1 km (Myrberg et al., 1976).

Well more than 25,000 fish species are in existence today, more than all other vertebrate species combined. The acoustic behavior of perhaps 100 of these species, representing only 0.4 percent, is known to some extent.

Fish produce sounds by a variety of mechanisms. Many of these involve striking two bony structures against one another. The swim bladder, an organ located in the abdominal cavity of most fish that contains air and regulates buoyancy, amplifies the fundamental frequency and matches the impedance of the sound to water (see Glossary for definitions of specific acoustic impedance and characteristic impedance). As a result, sounds produced by fish are pulsed signals with the energy mostly below 1 kHz. The pulses may contain broadband sounds if they are produced when two

---

Box 2-2
Deep, Dark, and Noisy?—Lantern Fish

It is likely that many more aquatic species produce and use sounds than currently documented. Indeed, this suggestion is supported by observations on a range of species showing that many have muscles or other structures similar to those known to be used for sound production in other species. One important example of this are the observations of Marshall (e.g., 1962, 1967), who showed that deep-sea fish of the family *Myctophidae* (lantern fish) have muscles that are connected to the swim bladder, which Marshall suggested are for sound production. Indirect support for such an argument comes from studies of the ears of these species showing that they have highly specialized sound detection systems (Popper, 1980), which could presumably have coevolved with evolution of sound communication. Sounds produced by myctophids may have direct relevance for some marine mammals, since it has recently been shown that these species are a direct part of the food chain for at least one species of *Stenella*. Although lantern fish make up perhaps the largest portion of fish biomass, their possible use of sound remains speculative at present.

bones strike one another, or a fundamental frequency and its harmonics when the sounds are produced by a muscle that is amplified by the swim bladder (see reviews by Tavolga, 1971; Demski et al., 1973; Myrberg, 1981; Zelick et al., 1999).

The overall contribution of fish sounds to the ocean noise budget has not been quantified. However, the character of fish sounds in some specific environments has been studied. For example, those in coastal shallow water and coral reef regions off the East Coast (e.g., Loye and Proudfoot, 1946; Fish, 1964; Fish and Mowbray, 1970) and West Coast (e.g., Johnson, 1948; Knudsen et al., 1948; Wenz, 1964; D'Spain et al., 1997) of the United States and around Australia (e.g., Cato, 1978, 1980; McCauley, 2001; McCauley and Cato, 2001). The degree to which these sounds are present varies from one ecosystem to the next and on diurnal and seasonal timescales (Tavolga, 1964). The major contributions are from those species that participate in chorusing behavior. Biological choruses occur when a large number of animals are calling simultaneously. Fish choruses are known to increase the ambient noise levels in certain locations, at certain times of the day, for example, the "sunset chorus" that lasts for a few hours after sundown and at certain times of the year (often the spring and early summer months) by 20 dB or more in the 50-Hz to 5-kHz band over sustained periods of time (see references listed above). Choruses appear to play an important role in spawning behavior in many species (Winn, 1964; Sancho et al., 2000a, b) and may be used by males to attract females to spawning sites (Winn, 1964; Holt, 2002).

While less recognized as sound producers than fish or marine mammals, a number of marine invertebrates produce sound. Some of these species produce choruses with a diurnal variation similar to those of soniferous fish (Fish, 1964). However, the sounds from the best-known sound-producing invertebrate, the snapping shrimp, display little diurnal variability. These animals are from a variety of species of the genera *Alpheus* and *Synalpheus*. They generate high levels of sound in the process of creating a focused jet of water by snapping closed their one large major frontal chela (fighting claw) (UC Division of War Research, 1946). The jet of water exits from the chela so quickly that the water is torn apart at the tail of the jet, referred to as cavitation (see Glossary). The subsequent collapse of the surrounding water into the void left by the jet is the source of the snapping sound (Versluis et al., 2000). The jet of water is sufficiently powerful to break standard aquarium glass and is believed to be used mainly for fighting and defense and for stunning and killing prey.

Colonies of one species of snapping shrimp (*Synalpheus regalis*) that dwell in the interstices of sponges in Caribbean coral reefs recently were discovered as the first marine animals to display eusocial behavior (Duffy, 1996). Eusocial behavior is a highly evolved, cooperative breeding behavior where each colony is centered around a single reproductive female, much like a beehive or an anthill. A loose analogy can be made between a snapping shrimp's water jet and the stinger of a bee. The sound produced in the process of creating a jet of water appears only to be a byproduct; no evidence suggests these sounds are used for communication or that snapping shrimp can detect sounds. However, the significant background noises produced by snapping shrimp are known to result in bottlenose dolphins changing the frequency of their echolocation clicks to move them outside the bandwidth of snapping shrimp noise, presumably to prevent the shrimp noise from masking detection of lower-frequency echolocation clicks (e.g., Au, 1993).

Much of the work on the worldwide distribution and underwater acoustics of snapping shrimp was done during World War II because of the impact of these sounds on the performance of military sonars. Results were published in the late 1940s (UC Division of War Research, 1946; Everest et al., 1948; Johnson, 1948). Snapping shrimp are found in shallow (less than 60 m), warm (greater than 11°C year-round) waters between 40° N and 40° S latitudes on stationary, rough ocean bottoms such as those covered by rocks or shells, in coral reefs, along pier pilings and jetties, and other areas where they can be protected (UC Division of War Research, 1946). The spectra of underwater acoustic measurements collected in the vicinity of snapping shrimp colonies show broad peaks in the 2-15 kHz band. Recent work extending the measurements to frequencies above the sonic band (Cato and Bell, 1992; Cato, 1992; Au and Banks, 1998) showed that snapping shrimp sounds contain energy up to 200 kHz and that individual snaps can have peak-to-peak source levels as great as 189 dB re 1 µPa at 1 m. Some additional places where recent studies of snapping shrimp sounds have been conducted are Gladstone, Queensland (Readhead, 1997), San Diego Bay (Epifanio et al., 1999), and Sydney Harbor (Ferguson and Cleary, 2001).

ANTHROPOGENIC CONTRIBUTIONS TO MARINE NOISE

Whether intentional or unintentional, human activity generates noise in the marine environment, and it is an important component of the total oceanic acoustic background. Sound is an important tool and a byproduct of a broad range of marine activities. To catalogue anthropogenic sound sources with their spatial and temporal variability and acoustic source characteristics they have been grouped into six categories: shipping, seismic surveying, sonars, explosions, industrial activity, and miscellaneous.

The extreme range of values of time, space, and signal structure variability make generalizations necessary (Table 2-1).

Vessel Traffic

Especially at low frequencies between 5 and 500 Hz, vessel traffic is a major contributor to noise in the world's oceans. Distant traffic contributes to the general acoustic environment in this frequency range; very large geographic areas are affected. In distant traffic noise, individual vessels are spatially indiscernible and often indistinguishable by frequency or temporal characteristics. Low-frequency ship noise sources include propeller noise (cavitation, cavitation modulation at blade passage frequency and harmonics, unsteady propeller blade passage forces), propulsion machinery such as diesel engines, gears, and major auxiliaries such as diesel generators (Ross, 1976). Particular vessels produce unique noise source levels with frequency, known as acoustic signatures. Sharp peaks (tones) produced by rotating and reciprocating machinery such as diesel engines, diesel generators, pumps, fans, blowers, hydraulic power plants, and other auxiliaries can be seen in the acoustic signature of a merchant vessel (Figure 2-1). Propeller blade passage tones and their harmonics, as well as propeller blade rate modulation of propeller cavitation, also contribute to the tonal structure of typical ship signatures and are particularly evident at lower ship speeds. With increased ship speed, broadband noise-generating mechanisms, such as propeller cavitation and hydrodynamic flow over the hull and hull appendages, become more important, essentially "masking" the machinery-

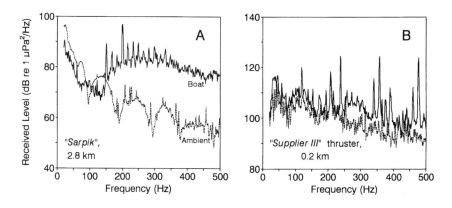

FIGURE 2-1 Received underwater sound spectral densities for two diesel-powered boats: (a) *Imperial Sarpik* at range 2.8 km, and (b) *Canmar Supplier III* with 336 kW (450 hp) bow thrusters at 0.2 km. The dotted spectrum is ambient noise before or after boat measurement. Note the different vertical scales in (a) and (b). Reanalyzed from recordings of Greene (1985); analysis bandwidth 1.7 Hz. SOURCE: Richardson et al., 1995, courtesy of Academic Press.

related tones observed at lower speeds. These spectral characteristics of individual ships and boats can be observed at relatively short ranges and in isolated environments. At distant points, multiple vessels contribute to the background, and it is this superposition of many distant sources that is characterized by broad spectral peaks labeled "usual traffic noise" in the Wenz curves (see Plate 1).

Globally, commercial shipping is not uniformly distributed. The major lanes are great circle routes (unless they extend to very high latitudes) or follow coastlines to minimize the time at sea. Dozens of major ports and several "megaports" handle the majority of the traffic, but in addition there are hundreds of small harbors and ports that host some level of daily seagoing traffic. The U.S. Navy's Space and Naval Warfare Systems Command defines 521 ports and 3,762 traffic lanes in its efforts to catalogue commercial and transportation marine traffic (Emery et al., 2001).

Other vessels may be found in widely distributed areas of the oceans outside of ports and shipping lanes. These include military craft in fleet exercises, fishing vessels, single vessels such as scientific research ships in a specific location on a one-time basis for measurements, and recreational craft typically near shore.

The contribution from recreational boating to the underwater noise field has not been quantified. Much of this boating activity occurs in shallow coastal waters, environments that are inhabited by many marine mammal species. Information on one aspect of the issue can be obtained from the National Marine Manufacturers Association, which publishes statistics on the number of U.S. boat registrations by state per year and the numbers of boats in various categories (outboard, inboard, sterndrive, personal watercraft, sailboats, and miscellaneous) owned in the United States in a given year (National Marine Manufacturers Association, 2002). For example, the number of boats owned in the United States increased from 15.8 million in 1995 to nearly 17 million in 2001, representing more than a 7 percent increase. Additional information on personal watercraft, a subset of the recreational boating sector, can be obtained from the Personal Watercraft Industry Association (2002). Measurements of the radiated noise from these watercraft are reported, but they pertain to the atmospheric radiated noise because of the potential impact on human coastal communities. Concern for this human impact has led the personal watercraft manufacturers to reduce atmospheric radiated noise levels by 70 percent since 1998. Many of the noise reduction techniques probably also have resulted in a decrease in underwater radiated noise levels. However, some of this 70 percent reduction has been achieved by rerouting the engine exhaust from above the water line to below, so that the overall change in underwater noise is difficult to predict.

Vessel operation statistics are complex to derive because of different criteria for defining ship type in different databases. Indeed, depending on how different analyses are done, even a single database, such as that produced by Lloyd's of London, can provide markedly different numbers of ships in the same category. The data mined for Table 2-2 show an increase of the commercial fleet from 72,662 in 1995 to 81,867 in 1999, an increase of 12 percent over four years. The trends all indicate growth consistent with population growth and use of the sea for economic, recreational, and transportation purposes. Economic pressure for oceanic shipping remains strong, and there is no near-term alternative available to move the necessary tonnage of goods and material globally. International economic infrastructure results in more raw materials being exchanged in the trade process. Fishing vessels account for approximately 23,000, or 28 percent of the world fleet. Bulk dry and oil tankers represent nearly 50 percent of the total tonnage but less than 8 percent of the vessel count.

Noise from Individual Ships and Boats

Databases of radiated noise measurements exist for some classes of surface ships. The largest collection of deep-water merchant ship radiated noise measurements probably is the Lloyd's Registry of London database (Lloyd's Registry of Shipping; see also Wales and Heitmeyer, 2002). However, limited information is readily available regarding the acoustic signatures of some of the types of commercial vessels listed in Table 2-2. These data often are found in technical memoranda, databases that require international data exchange agreements (e.g., Lloyd's Registry of Shipping), Navy-related databases, and other sources not readily available to the public and research community.

Every vessel has a unique signature (Figure 2-1), which changes with ship speed, the condition of the vessel, vessel load, the activities taking place on the vessel, and even with the properties of the water through which the ship is traveling (Ross, 1976). However, high-quality shipping noise modeling probably requires only representative source spectra for the different classes of ships (Figure 2-2). Source spectral densities for the five classes of surface ships are used in the ANDES (Ambient Noise Directionality Estimation System) (Renner, 1986a, b; 1988) as well as the RANDI (Research Ambient Noise Directionality) (Wagstaff, 1973; Hamson and Wagstaff, 1983; Schreiner, 1990; Breeding, 1993) models. The curves for the two models differ according to the way the various classes are defined and the modeling approach taken; the levels in ANDES depend solely on the class of ship, whereas ship length and ship speed are used to calculate a scaling factor based on empirically derived power laws in the RANDI model. (The ANDES source spectral densities also are used in the newly developed Dynamic Ambient Noise Prediction System; see Chapter 4). Using the RANDI model, source spectral density levels range from more than 195 dB re 1 $\mu Pa^2/Hz$ at 1 m around 30 Hz for fast-moving, large supertankers

TABLE 2-2 Principal Commercial Fleets of the World—by Ship Type Category (GT = gross tons)

Ship Type	1995 No.	1995 GT	1997 No.	1997 GT	1999 No.	1999 GT
Liquefied natural gas tanker	91	7,091,934	103	8,298,330	113	9,280,153
Liquefied petroleum gas tanker	894	7,807,087	942	8,227,819	978	8,649,191
Chemical	2,077	12,073,051	2,260	13,643,913	2,456	16,310,943
Crude oil tanker	1,656	118,835,028	1,717	122,377,669	1,782	127,875,958
Oil products tanker	5,105	24,685,537	5,216	24,729,866	5,269	26,215,579
Other liquids	315	415,793	347	598,154	343	565,418
Total	10,138	170,908,430	10,585	177,875,751	10,941	188,897,242
Bulk dry	4,799	128,517,859	5,079	140,921,192	4,881	139,408,629
Bulk dry/oil	226	14,105,815	242	11,420,914	219	9,562,092
Self-discharging bulk dry	158	2,922,535	157	2,953,916	164	3,170,689
Other bulk dry	982	6,148,064	1,074	6,872,729	1,093	6,816,236
Total	6,165	151,694,273	6,552	162,168,751	6,357	158,957,646
General cargo	17,161	56,739,241	17,467	56,569,318	16,880	55,981,408
Passenger/general cargo	351	675,939	342	602,641	348	609,892
Container	1,763	38,742,105	2,187	48,839,028	2,457	55,255,401
Refrigerated cargo	1,466	7,158,402	1,443	7,145,501	7,415	7,037,866

Roll on/roll off cargo	1,673	20,429,523	1,742	21,978,592	1,844	25,256,499
Passenger roll on/roll off cargo	2,256	1,562,021	2,425	12,121,320	2,553	12,824,778
Passenger (cruise)	287	4,979,116	328	5,877,376	345	7,194,096
Passenger ship	2,236	1,190,7921	2,500	1,239,630	2,595	1,334,411
Other dry cargo	216	1,885,770	259	1,993,038	267	2,045,037
Total	27,409	144,080,038	28,693	156,366,444	34,704	167,539,388
Fish catching	23,111	11,005,206	22,729	10,647,509	23,003	10,613,938
Other fishing	818	2,342,715	811	2,024,838	838	1,636,874
Total	23,929	13,347,921	23,540	12,672,347	23,841	12,250,812
Offshore supply	2,382	1,869,241	2,370	1,960,707	2,528	2,315,487
Other offshore	463	2,492,073	554	3,016,539	611	4,563,628
Total	2,845	4,361,314	2,924	4,977,246	3,139	6,879,115
Research	818	1,106,990	618	1,112,785	846	1,271,695
Towing/pushing	7,721	2,085,277	8,603	2,275,408	9,044	2,401,953
Dredging	1,125	1,874,095	1,120	1,931,933	1,121	2,279,743
Other activities	2,650	2,897,862	2,659	2,746,528	2,824	2,929,967
Total	12,314	7,964,224	13,000	8,066,654	13,835	8,883,358
World totals	72,662	321,447,770	74,709	344,251,442	81,876	354,510,319

SOURCE: Lloyd's Register, Fairplay Ltd., World Fleet Statistics, 2001.

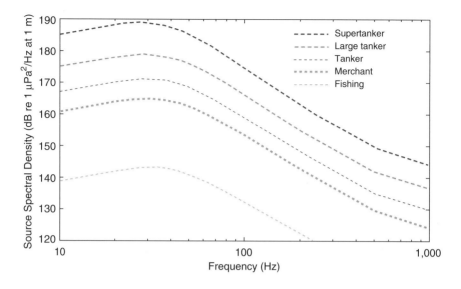

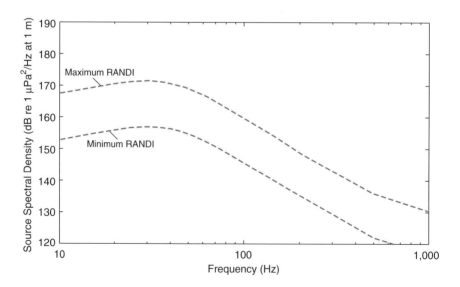

FIGURE 2-2 (a) Modeled surface ship source spectral densities for the five classes of ships used in the RANDI ambient noise model. The curves in each class also are a function of ship length and ship speed; those plotted in the figure pertain to the mean values of ship length and ship speed in each class. (b) A comparison of the maximum and minimum merchant ship source spectral densities from the RANDI model (calculated using the maximum and minimum ship lengths and ship speeds for this class; re Table 2-3). SOURCE: Wagstaff, 1973. Adapted from data from the Naval Undersea Center.

down to 140 dB re 1 $\mu Pa^2/Hz$ or less for smaller craft such as fishing vessels (Table 2-3).

Figure 2-2b also shows a comparison of the "merchant" class source spectral densities in RANDI with the mean source spectral density in Wales and Heitmeyer (2002) calculated as the decibel mean over 54 merchant-class source spectral densities. The model of the acoustic source used in Wales and Heitmeyer to derive the source spectral densities from the measured received spectral densities is a vertical line of incoherent point sources, rather than a single-point source, in order to more accurately account for the character of the acoustic source region about the ship propeller. An interesting observation is that the decrease in ship spectral density levels with frequency above 400 Hz has the 5-6 dB/octave dependence as seen in the Knudsen curves for wind-generated noise (see Plate 1). Wales and Heitmeyer also analyzed the variations of individual merchant ship spectra from their mean spectrum; variations are significantly greater below 400 Hz (up to 5.3 dB standard deviation) than above (a standard deviation of about 3.1 dB).

As mentioned previously, ship-generated spectra are composed of a broadband component, predominantly the result of propeller cavitation, and a set of harmonically related tones created both by propeller cavitation (the blade lines) and the machinery on the ship. The broadband and tonal components produced by cavitation account for 80-85 percent of ship-radiated noise power (Ross, 1976). The discussion above pertains to the character of the broadband component. Source-level models also have been developed for the propeller fundamental blade rate line occurring predominantly in the 6-10-Hz band for the world's merchant fleet (Gray and Greeley, 1980).

Acoustic signature data are available for some oceanographic research ships and boats. Although they may be important locally, noise levels are typically so low they are unimportant in the general acoustic environment of the world's oceans. There is a significant literature dealing with the effects of fishing vessel noise on fish populations on which marine mammals may depend (Mitson, 1995). Observed responses of marine mammals to boats, not just fishing boats, are discussed in Chapter 3. Signature data are not readily available for most survey or observation vessels, although some examples are presented in Richardson et al. (1995) and as unpublished documents and reports. A sampling of whale-watching boat signatures is available in the published literature (e.g., Richardson et al., 1995; Erbe, 2002).

Extensive acoustic signature data exist for military surface ships. Individual vessel signature data resources are classified and held in government agencies such as the Naval Research Laboratory and cannot be used in this study. The U.S. Navy does post vessel descriptions as well as current deployment numbers on its Web site.

TABLE 2-3 Source Spectral Densities for Commercial Vessels Underway for Several Frequencies

Ship Type	Length (m)	Speed (m/s)	Source spectral density (dB re 1 $\mu Pa^2/Hz$ at 1 m)					
			10 Hz	25 Hz	50 Hz	100 Hz	300 Hz
Supertanker	244-366	7.7-11.3	185	189	185	175	157
Large tanker	153-214	7.7-9.3	175	179	176	166	149
Tanker	122-153	6.2-8.2	167	171	169	159	143
Merchant	84-122	5.1-7.7	161	165	163	154	137
Fishing	15-46	3.6-5.1	139	143	141	132	117

SOURCE: Adapted from Research Ambient Noise Directionality (model) (RANDI) source-level model in Emery et al. (2001) and Mazzuca (2001).

Pleasure craft do not contribute significantly to the global ocean acoustic environment but may be important local sources of underwater noise. High-speed ocean yachts are expected to be sources of high noise levels but are sufficiently small in number as to represent significant sources only local to the individual craft. Results from a recent study of source signatures from outboard, inboard-outboard, and inboard powerboats shows that source levels for the largest amplitude narrowband tones typically range between 150 and 165 dB re 1 μPa at 1 m and the broadband radiated energy, which is engine RPM dependent, has maximum source spectral density levels in the 350-1,200-Hz band of 145-150 dB 1 $\mu Pa^2/Hz$ (Bartlett and Wilson, 2002). Additional examples of individual ship signatures in these classes can be found in Richardson et al. (1995).

Future Trends in Shipping

Although the number of vessels and tonnage of goods shipped are increasing (e.g., a nearly 30 percent increase in volume shipped by the U.S. fleet over the past 20 years; 1,793.9 million metric tons [mmt] in 1980 to 2,331.6 mmt in 2000) (U.S. Maritime Administration, 2002), the relative distribution of numbers of ships among the various classes is not expected to change remarkably in the future. If dramatic changes are made to the shipping fleet, they likely will be mandated by economic forces such as more efficient or cheaper propulsion systems, faster ships, or hull configurations that allow more bulk tonnage. Any one of these changes could have a significant impact on a ship's radiated noise characteristics. A discussion of the long-term trends in shipping contributions to ocean noise is presented later in this chapter.

Marine Noise Generated by Oil and Gas Industry Activities

Oil and gas industry activities may be divided into five major categories: (1) seismic surveying, (2) drilling, (3) offshore structure emplacement, (4) offshore structure removal, and (5) production and related activities (including helicopter and boat activity for providing supplies to the drilling rigs and platforms). Offshore seismic surveying, the predominant marine geophysical surveying technique employed by the oil and gas industry, uses intentionally created sound. The last four activities listed create primarily unintentional noise and will be discussed in less detail. The noise levels associated with oil and gas production are typically much lower than those involved in seismic surveying (see Richardson et al., 1995).

Offshore oil industry activities have a patchy distribution along the world's coastlines, ranging from about 72° N latitude to about 45° S latitude. Seismic surveying activity and oil and gas production have taken place off the coasts of North and South America, Africa, Europe, Asia, and

Australia. Activity levels associated with well drilling and seismic surveying by the oil and gas industry are monitored by various industry trade and database companies and published on a monthly basis in various trade journals, such as *Hart's E&P*, *Journal of Petroleum Technology*, *Oil and Gas Journal*, *Offshore*, and *World Oil*. The companies actually performing the work provide the numbers about these activity levels to the database companies.[1]

Seismic Reflection Profiling

Seismic reflection profiling encompasses a variety of methods, all of which use sound to relay information about geological structure beneath the surface of the earth. The oil industry relies on the extensive use of seismic reflection profiling to provide unique information about the rocks that extend beneath the seafloor, down to depths exceeding 10 km. Seismic reflection profiling, which includes what is commonly called three-dimensional (3D) seismic, is also used by academic and government groups, as well as the mining, environmental consulting, and other industries, to gather information about subsurface rock properties. The major operational elements in industrial marine 3D seismic reflection surveying are (1) the seismic vessel, typically about 100 m long by 30 m wide; (2) one or two air-gun arrays towed about 200 m behind the seismic vessel; and (3) cables, called streamers, containing large numbers (on the order of a few thousand) of hydrophone sensors towed behind the seismic vessel. Current technology uses streamers up to 12 km long to record the echoes returning from the subsurface (Figure 2-3). In the open marine environment, air-guns are the most commonly used sound source, but explosives buried in drilled holes are used to acquire similar data in waters shallower than about 4 m.

Marine seismic reflection profiling currently relies on the use of arrays of air-guns. These arrays have replaced the explosive charges that previously were used as sources.[2] Air-guns release a volume of air under high pressure, creating a sound pressure wave that is capable of penetrating the

---

[1] The numbers provided by one database group may differ from those provided by another group because of the use of different categories of activity (e.g., under contract versus actually working) and because of different ways of reporting requested by the database companies of the different contractors. This can make reconciling numbers from one group's report to another's report difficult, so that a single database company's numbers should be used to develop industry trend information.

[2] Explosions still are employed in a few government and research-related ocean-going experiments, as are seismic waterguns. Waterguns do not release air as part of the pulse generation process in order to maximize the signal pulse-to-bubble pulse ratio. However, they are not generally used since they are more inefficient and their signature contains higher frequencies than an air-gun.

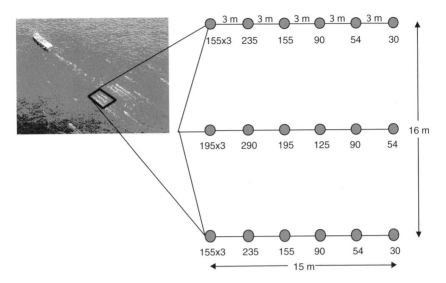

FIGURE 2-3 Schematic diagram of an air-gun array. A total volume of 3,397 cu. in. is shown. This array has three subarrays (each line of circles) and uses 24 air-guns. Each circle represents an air-gun, except for the circles at the head of each array, which represent three-gun clusters. The nearest number represents the volume of air expelled by individual air-guns in cubic inches.

seafloor to determine substrata structure. Each complete air-gun array used in the seismic industry will typically involve 12-48 individual guns. Most of the seismic industry uses air-gun arrays with operating pressures of 2,000 psi (equal to 13.8 million pascals) and are typically about 20 m by 20 m.

The acoustic pressure output of an air-gun array is (1) directly proportional to its operating pressure; (2) directly proportional to the number of air-guns, all else being equal; and (3) proportional to the cube root of the volume. For example, an 8,000-cu.-in. (0.131 m³) array has a 3.4-dB greater output than a 2,500-cu.-in. (0.041 m³) array having the same number of guns.[3] The acoustic pressure signal of air-gun arrays is focused vertically, being 12-15 dB stronger or more in the vertical direction for some arrays in use today. The ability to focus the sound output in the vertical direction is a function of the total array aperture in both the fore-and-aft and side-to-side directions and the number of air-guns in the array

---

[3] $= 20\log_{10} (8000/2500)^{1/3}$, if the difference in each single-gun volume is also in the same 8,000:2,500 ratio. A 48-gun array has about a 12.0-dB greater output than a 12-gun array, *almost* regardless of the total volume of the array [$= 20\log_{10}(48/12)$].

(Plate 4). Vertical output can be maximized while minimizing output in the horizontal plane through the use of arrays incorporating more small air-guns rather than fewer larger air-guns.

The literature, including both that published by the seismic exploration industry and by bioacousticians, refers to back-calculated levels of up to 260 dB re 1 μPa at 1 m for the maximum output pressure levels [zero-to-peak, subtract about 10 dB to obtain root mean squared (RMS) value, per W. J. Richardson, personal communication, 2002] of industry air-gun arrays (Richardson et al., 1995; Dragoset, 1990). The back-calculation is valid for point sources, not ones that measure 20 m on a side, so that the 260 dB should be used to calculate sound pressures in the vertical far field. The far-field distance is a function of the array dimensions, the speed of sound in water, and the frequency of the source. The maximum pressure level an animal could experience from an air-gun source in use today in the seismic industry will be in the range of 235-240 dB re 1 μPa (RMS). The location where this level of sound is attained will be vertically beneath the air-gun array, generally near its center, but the exact location and depth beneath the array are dependent on the detailed makeup of the array, the water depth in which the array is operating, and the physical properties of the seafloor above which the array is operating (Dragoset, 2000).

The peak amplitude of an air-gun array is also a function of the frequency (Figure 2-4). The peak pressure levels emitted from commonly used seismic industry air-gun arrays are in the 5-300 Hz range. The guns are towed at a speed around 5 knots (2.6 m/s) and are fired about every 10-12 seconds. A typical seismic operation includes a series of parallel passes by a vessel towing one or two air-gun arrays and 6-10 streamers. Turning typically takes about two hours, and the air-guns are shut down during this maneuver. In addition to this turning period, the air-guns do not operate when the vessel is in transit to and from the survey site, when sufficiently bad weather occurs, when streamers are being deployed or retrieved, or when critical equipment fails. Given these constraints, air-guns are generally firing less than 40 percent of the time the vessel is underway (Philip Fontana, personal communication, 2002).

Marine seismic crews are much more efficient today than they were 10 years ago, since more and longer streamers are towed now than in the past (DeLuca, 2000; Eng, 2001; Maksoud, 2001). The acquisition footprint (0.25 times the total length of the streamer times the total distance from the starboardmost streamer to the portmost streamer) can be as much as 4.24 km². In other words, a seismic crew can get into and out of a specific area much more quickly than in times past because fewer tracks are required, given the wider coverage (swath) per track. The use of seismic time-lapse monitoring for reservoir management (repeating seismic surveys to monitor changes in a hydrocarbon reservoir over months and years) means that

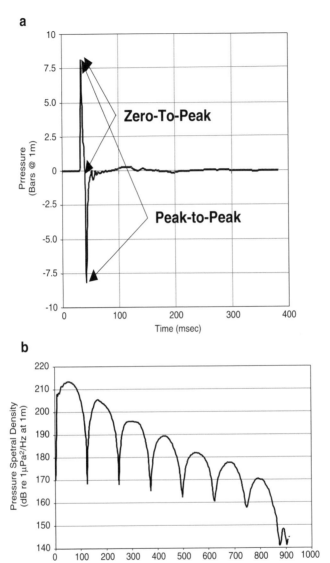

FIGURE 2-4 Acoustic signal of a 4,550 cu. in. air-gun. (a) Typical pulse created by the firing of an air-gun array. The high-amplitude portion of the pulse lasts about 20 ms. (b) An amplitude spectrum of an air-gun signal. This plot shows pressure levels as a function of frequency for a signal generated by a 4,550-cu.-in. air-gun array. Courtesy of Philip Fontana, Veritas DGC.

more seismic surveys are likely to be shot over producing fields than was true in the past.

Noise Generated by Other Hydrocarbon Industry Activities

Drilling techniques employed by the oil and gas industry require a wide variety of equipment (Box 2-3). At any given time, it can be assumed that representatives of each of these types are in use somewhere in the world (Table 2-4). When drilling is taking place, auxiliary noise is generated, created by activities including supply boat and support-helicopter movements. The worldwide offshore mobile rig count can vary over time as a result of business conditions in the oil industry (Figure 2-5). A comparison of rig counts (Table 2-4, Figure 2-5) highlights the differences in reporting from the groups. These graphics illustrate the overall numbers of rigs of all types operating at a given time, an idea of year-to-year variability, and a current distribution of rigs for different areas around the world.

Jack-ups are the most commonly used offshore drilling equipment, followed closely by the use of platform rigs (see the Offshore Rig Locator published monthly by the ODS-Petrodata Group). The sound pressure levels created by the different drilling methods are not well known. Richardson et al. (1995) present a small amount of data, mostly recorded from the monitoring of projects along the North Slope of Alaska and the adjoining coast of Canada. In general, drill ships are the noisiest type of drilling equipment being used, with a maximum broadband source pressure level across the 10-Hz to 10-kHz band of about 190 dB re 1 μPa at 1 m (RMS) (Richardson et al., 1995). Drill ships are expected to be the noisiest

---

Box 2-3
Oil and Gas Extraction Platforms

Platform rigs: permanently mounted rigs located on stationary production structures

Semisubmersibles: mobile, steel-decked structures whose hollow support structures do not rest on the seafloor

Jack-ups and submersibles: mobile, steel-decked structures whose legs or other support structures rest on the seafloor

Drill ships: ships with drilling capabilities

Drill islands: artificial islands upon which drilling rigs are placed, constructed in areas normally covered by ice substantial portions of the year

TABLE 2-4 International Offshore Mobile Working Rig Count

Location	January 2002	December 2001	January 2001
Canada	7	7	6
Europe	71	61	53
Middle East	37	37	30
Africa	20	18	23
Latin America	46	47	45
Asia Pacific	61	59	55
United States	126	123	174
Total	368	352	386

SOURCE: Hart's E&P, April 2002.

because the hull is an efficient transmitter of all of the ship's internal noises, and the ships do not anchor but use thrusters to remain on location, resulting in propeller noise much of the time during the drilling operation. Research is needed to make accurate measurements of the sound pressure levels generated by various drilling techniques.

The compilation of drilling activity numbers over time with a conventional geographic breakdown as illustrated by Table 2-4 may not be particularly useful to describe the drilling ensonification of the oceans. Neither changes in relative percentages of the different drilling technologies being used nor changes in the distribution of activities between shallow water and

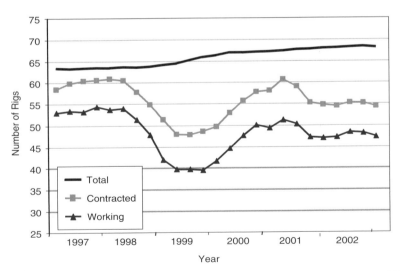

FIGURE 2-5 Worldwide offshore mobile rig numbers from 1997 to present. This figure does not include rigs permanently located on platforms, of which there were 139 contracted as of March 4, 2003. SOURCE: ODS-Petrodata, Houston.

deep water are reflected in general oil rig estimates. Rapid changes in drilling technology and equipment are likely to change the noise generated and are not included. Sound pressure level measurements are needed to conclude how these changes in oil industry techniques affect ocean noise.

Offshore structure emplacement will create some localized unintentional noise for relatively brief periods of time. Because a few large structures that will span relatively great water depths are emplaced each year, extremely powerful vessels are required to transport them from the point of fabrication to the point of emplacement. This activity lasts for a few weeks and currently does not occur more often than 8–10 times per year. The installation of subsea structures, primarily in deep-water sites, is becoming more commonplace. Measurements of the sound pressure levels associated with such activity have yet to be made.

Additional noise is generated during oil production activities, which can include borehole logging, casing, cementing, perforating, pumping, pipe laying, pile driving, ship and helicopter support, and others to support rig and platform work. Impulsive hammering sounds created by installation of conductor pipe resulted in received sound levels of 131-135 dB re 1 μPa recorded 1 km from the source (see Richardson et al., 1995). Assuming transmission loss resulting from spherical spreading, this will translate to 195 dB re 1 μPa at 1 m, with the peak amplitudes occurring at around 40 and 100 Hz.

Oil Industry Noise—Future Trends

Oil and gas industry activities occur along many continental margins between 72° N latitude and 45° S latitude. The major areas of current oil industry activity include northern Alaska and extreme northwest Canada, the east coast of Canada, U.S. Gulf of Mexico, Mexican Gulf of Mexico, offshore Venezuela, offshore Brazil, offshore West Africa, offshore South Africa, North Sea (most sectors), Middle East, northwest Australia, New Zealand, southern China, Vietnam, Malaysia, and Indonesia. It is unlikely that any major shifts in gross patterns of where this industry operates will take place in the near future.

Noise levels associated with new exploration activities may be more noticeable when they occur in relatively quiet areas with little human activity. Recent new exploration areas include the deep-water U.S. Gulf of Mexico and deep-water offshore of West Africa, both of which basically have become active in the past 5-10 years. Local and global economic changes can abruptly modify exploration plans, so noise levels that are affected by exploration activities can change rapidly. How quickly drilling activity is turned off and on depends in large part on which specific oil companies are operating in a particular geologic province or basin. How fast seismic surveying activity is turned off and on depends on global economic factors, as well as local economic conditions, and again on which specific oil companies are in a given area.

Oil and gas production is moving into water depths up to 3,000 m. Those depths will require the use of drill ships and most likely require the use of floating production, storage, and off-loading systems that involve the use of oil tankers, most likely on nearly a weekly basis, since high production rates are needed to justify the expense of deep-water fields. Therefore, the deep-water fields may be a source of greater noise than the shallow-water fields have traditionally been, at least in terms of base noise levels. The Minerals Management Service in the United States is now recognizing this as a growing issue.

Sonars

Sonar (sound, navigation, and ranging) systems use acoustic energy to characterize physical properties and locate objects beneath the ocean surface. The wide range of applications requires systems that vary greatly in engineering specifications and deployment strategies. Sonar systems have both military and civilian applications and can be divided into low-frequency (<1 kHz), mid-frequency (between 1 and 10 kHz), and high-frequency (>10 kHz). Generally, military systems exist for all frequency ranges, while civilian systems are confined to the higher frequencies.

Military Sonars

Military sonars are typically operated at higher power levels than civilian sonars and are used for target detection, localization, and classification. Military low-frequency sonars are used for surveillance and are designed to gather information over large areas. If conditions permit, these sonars can collect information over entire ocean basins. The mid-frequency military systems are tactical sonars and are designed to look over tens of kilometers for the localization and tracking of targets. Military high-frequency systems are either weapons (torpedoes and mines) or counterweapons (mine countermeasure systems or antitorpedo devices) and are designed to perform over hundreds of meters to a few kilometers.

Active sonar technology, currently deployed by the navies of the world and undergoing further development, seeks targets by sending out high-energy acoustic pulses and recapturing the echo. Characteristics of the echo provide information on the objects that reflect and scatter the pulses. The class of surveillance sonars presently in the fleet is designed to locate targets, primarily submarines and to some extent surface ships, at tens to hundreds of kilometers away to provide early alerts of potential threats to navy vessels. The U.S. Navy's Surveillance Towed Array Sensor System Low Frequency Active (SURTASS-LFA) system utilizes a vertical line array of up to 18 source projectors operating in the frequency range of 100-500 Hz. The source level of each projector is approximately 215 dB re 1 μPa at 1 m (Johnson, 2001). In addition, the U.S. Navy reports that the hull-mounted AN/SQS-53C tactical sonars can generate pulses in the 1-5 kHz band and have been operated at source levels of 235 dB re 1 μPa at 1 m and that the AN/SQS-56 sonars generate pulses in the 5-10 kHz band and have operated at 223 dB re 1 μPa at 1 m source levels (Evans and England, 2001). High-frequency military sonars (above 10 kHz) are used for weapon and counterweapon location at ranges of tens to thousands of meters. The sonars are operated in a variety of modes, different signal types, and different signal lengths and strengths but typically over a relatively narrow frequency range. They can be used to generate broadband signals in which a wide range of frequencies are transmitted simultaneously, but it is not common to do so.

The use of military sonars typically is limited to operational areas, a small portion of the total ocean space (Lloyd's Register, 2001). If war situations are excluded, the remaining areas of military activity are well defined and the level of activity is also well defined and episodic; typical U.S. Navy individual ship exercises last a few hours to a few days, predeployment and battle group exercises normally last 10-12 days and involve a full carrier battle group or amphibious-ready group, and the duration of large multinational fleet exercises is up to a month, but these occur only every other year or so. The total number of military ships extant on the globe numbers in the thousands, but use of military sonar systems is limited to hundreds.

Other military active sonars include communication sonars for inter-platform information exchange, depth sounders, sidescan sonars for sea-floor mapping, and variants of the communication sonar that are used for device activation and event initiation, for example. Mine-hunting systems are high-frequency systems, ranging from tens of kilohertz for detection to hundreds of kilohertz for localization. The systems are highly directional and use pulsed signals.

Commercial Sonars

Commercial sonar systems are designed for special purposes such as depth sounding, fish finding, and obstacle detection. Typically, they operate at higher frequencies, project lower power, and have significant spatial resolution with narrower beam patterns and shorter pulse lengths than military sonars. Characteristics of the underwater transducers used in these commercial sonar systems, as well as in military sonar systems, can be obtained from the transducer manufacturers, including the International Transducer Corporation, Reson, and Massa Products Corporation. Commercial sonars typically operate in a narrow frequency band with a center frequency between 1 and 200 kHz or more, depending upon the application. The source levels of some of these sonar transducers can reach values of 250 dB re 1 μPa at 1 m (e.g., model TR-208A by Massa Products Corporation) (Massa Products Corp., 2002). These source levels are sufficiently high that sonar performance in shallow water becomes limited by cavitation (Urick, 1975).

Commercial depth sounders and fish finders typically are designed to focus sound in the downward direction, although forward-looking sonars also are available. A common type of fish finder/echosounder (e.g., model LS-6000 from Furuno) operates at two frequencies, 50 kHz and 200 kHz, with output power on the order of 1 kW (201 dB re 1 μPa at 1 m) at a duty cycle of 0.1 percent (0.2 ms pulse every 0.2 s). These frequencies are too high to be audible to fish; however, 50 kHz certainly falls within the range of hearing sensitivity of many marine mammal species. Even if only a small fraction of the 17 million boats owned in the United States in 2001 (National Marine Manufacturers Association, 2002) and the 80,000 vessels in the world's merchant fleet as of 1999 (Table 2-2) are equipped with commercial sonar systems, the potential exists for these systems to adversely impact the marine environment. Depth sounders typically operate in nearshore and shallow waters. However, fish finders are used in biologically productive areas in both deep and shallow waters that are likely to contain marine mammals.

According to the Pew Oceans Commission (2002), the mortality caused by the unintended capture during commercial fishing operations ("bycatch") exceeds sustainable levels for 13 of the 44 marine mammal species that suffer high mortality rates as a result of human activities. Low-power acoustic deterrent devices (ADD) are used in some fisheries in attempts to reduce this bycatch. Good evidence exists that these "pingers" are effective, at least for some marine mammal species. High-power ADDs (sometimes called acoustic harassment devices or AHDs) are designed to be sufficiently high level to exclude marine mammals, usually pinnipeds, from areas such as aquaculture sites, sections of river systems where migrating salmonids are vulnerable, and some fishing equipment. Whereas the low-power "pingers" have maximum source levels typically between 130 and 150 dB re 1 mPa at 1 m, the high-power AHDs have source levels in the 190-200 dB re 1 μPa at 1 m range. Both types generate a series of pulses, each lasting from 10 to 500 ms with interpulse periods ranging from negligible (i.e., continuously repeated transmissions) up to 10 s. The signals have frequency content in the 5-30-kHz band and some extend up to 160 kHz, which is sufficiently high to be outside the range of audibility of most species of fish.

In the United States, attention turned to the problem of bycatch more than a decade ago and resulted in an increase in the use of acoustic deterrents. Amendments to the Marine Mammal Protection Act during its reauthorization in 1994 included provisions to try to reduce marine mammal bycatch. U.S. fisheries now are placed in one of three categories:

I.   those with frequent serious injuries and death of marine mammals,
II. those with occasional serious injuries and death, and
III. those unlikely to cause serious injury or death.

The number of U.S.-registered vessels in each of these categories in each of the various U.S. fisheries is published at least yearly in the *Federal Register*. The most recent report (McKeen, 2002) shows about 14,300 vessels in category I, with 188 of these off the California and Oregon coasts, about 14,000 in the Atlantic, and 100 or so in the Gulf of Mexico and the Caribbean. The category II vessels number more than 25,300 with 7,364 off the West Coast of the United States, more than 17,950 in the Atlantic, and 50 in the Gulf of Mexico. Fishing boats that are in categories I and II must follow certain procedures, including registering yearly with the National Marine Fisheries Service for authorization to incidentally take marine mammals during fishing activities, allowing designated observers aboard when requested, and following "take reduction" plans developed for that fishery. Take reduction plans for some fisheries were put in place as early as 1997-1998, and some of these plans have included the deployment of low-power ADDs. As a result, the use of these pingers in the U.S. commercial fishing fleet has jumped in the past 4-5 years and continues to change with further development and modification of the various fisheries' take reduction plans.

Fishing nets with pingers exclude porpoises from their immediate vicinity. Although bycatch is thereby reduced, concerns have arisen that pingers could lead to significant habitat exclusion if used in sufficient numbers. In contrast, AHDs are sufficiently high level that they could actually damage the hearing of marine mammals exposed at close range. Pinnipeds that are highly motivated to prey on the fish being protected are particularly susceptible. In addition, AHDs used at aquaculture facilities have been shown to exclude nontargeted species, especially odontocetes. For example, Olesiuk et al. (2002) have shown that porpoise densities were significantly reduced when ADDs were in operation. No porpoises were found within 400 m of the device, and the sighting rates between 2.5 and 3.5 km were 10 percent of the control rates.

Underwater Sound Sources in Basic Ocean Acoustics Research

This section summarizes the characteristics of underwater sound sources used in basic ocean acoustic and acoustical oceanography research programs in the United States. Basic science programs in seismology that have seagoing experimental components involving the deployment of underwater seismic sources are not discussed here. Rather, the characteristics of seismic air-guns and air-gun arrays often used in these experiments are discussed under the Seismic Reflection Profiling section.[4] In addition, explosive charges used as seismic sources are discussed under the Explosive Sources section in this chapter. Active acoustic experiments, those involving acoustic signal transmissions, in advanced development programs for the operational navy also are not covered here because of their classified nature. They are best considered part of the operational navy activities discussed in the Military Sonars section.

Almost all of the basic ocean acoustics research programs in the United States are sponsored by the Office of Naval Research (ONR). Fewer than a dozen or so experiments are conducted each year and typically last one to three weeks. The sound sources in these experiments primarily are commercially available transducers, sometimes with small changes in design or deployment geometry, but also a few specially designed sonars are used to meet specific research objectives. In addition, sources often are rented from the Underwater Sound Reference Detachment (USRD). For example, a popular sound source from USRD that is used in low-frequency acoustics experiments is a Type J15-1 source. It is a moving coil-type device designed to transmit signals in the 30-900 Hz band with a maximum source level of approximately 170 dB re 1 µPa at 1 m (Ivey, 1991). A wide variety of waveforms are transmitted by these sonars over a wide range of frequency bands, source levels, and duty cycles because of the large number of research questions addressed in these programs. Explosive charges sometimes are deployed, although their use by the research community has been decreasing, partly because of safety and environmental concerns but also because of the lack of control over and the nonreproducible nature of the source waveform and the detonation depth. The spatial extent of the signals transmitted in most of these experiments, other than the basin-scale acoustic tomography experiments discussed below, are local in nature, less than a few tens of kilometers. Although the experiments have been conducted in various parts of the world, U.S. experiments typically occur in U.S. territorial waters.

The most widely known ocean acoustics research program is the Acoustic Thermometry of Ocean Climate (ATOC) program. It has received a high level of attention from regulatory agencies, the public, and the scientific community. The characteristics of the ATOC source signals have been documented in the two previous NRC reports on low-frequency

---

[4]The air-gun arrays used in seismic research typically have fewer and smaller guns than seismic industry arrays. An exception is the 20-element air-gun array operated by Lamont-Doherty Earth Observatory (LDEO) with 0-pk output source levels in the vertical direction of 260 dB re 1 µPa at 1 m according to the LDEO Web site.

sound (NRC, 1994, 2000). The source, deployed at a depth greater than 800 m, has a source level of 195 dB re 1 µPa at 1 m. Modern-day ATOC experiments and the few other recent basin-scale acoustic tomography experiments have spatial extents of hundreds of kilometers. These experiments are able to probe the properties of the ocean out to basin-scale ranges to thousands of kilometers because the transmitted signals are long and complex. Therefore, matched filtering of the received signal with the transmitted signal provides significant processing gain (the AG term in Equation 1-3).

Since the start of the ATOC program and the controversy surrounding it, ocean acoustic experimental programs have undergone an increasing degree of scrutiny. The ONR now requires all ONR-sponsored experimental research programs involving underwater acoustic transmissions to follow the planning process specified in the National Environmental Policy Act (NEPA). The purpose of this process is to clearly identify and mitigate against any potential effects of acoustic transmissions, as well as all other experimental activities, on the marine environment and to establish if an environmental impact statement must be prepared. All federal and relevant state environmental laws must be followed, including the Marine Mammal Protection Act, the Endangered Species Act, and the Magnuson-Stevens Fisheries Conservation and Management Act, as well as other laws, regulations, and executive orders.

Future Trends in Sonar Use

Commercial sonars will continue to proliferate in the oceans. Their acoustic characteristics will not change significantly, since absorption limits the frequency at the high end, physical size is a limitation at lower frequencies, and the properties of the water column itself at the very shallow depths of hull-mounted systems limit the maximum acoustic power output as a result of cavitation (Urick, 1975). However, acoustic power output is not a limiting factor for most commercial sonar applications in most environments. Potentially, research to increase the power of military sonars could continue, but operational requirements for higher power will require difficult research. Current materials and structural configurations cannot survive the extreme electric fields and displacements necessary to produce more acoustic power. An alternative approach is to deploy arrays of sources that circumvent these constraints but have limitations of their own, for example, they are difficult to deploy or difficult to maneuver while being deployed.

Perhaps the most uncertain are new applications for sonars. The ocean is still poorly understood, both in economic potential and in its role in the well-being of our planet. As terrestrial resources dwindle, the pressure to understand, map, and explore the ocean will rise dramatically. A variety of sonar devices provide the most effective methods, so research, development, and increasing use of new sonar tools are likely.

Sonars are designed and built worldwide—their characteristics are well understood and calibrated, and these data are generally available. The impacts on ocean organism populations are unknown. To what characteristics do marine mammals react—are they particularly susceptible to certain parameters such as rise time, power, signal length, particular frequencies, or rate of repetition? These characteristics may be modified to minimize effects on marine life without detracting from the purpose of the tool.

Explosive Sources

Explosive sources are broadband with very high zero-to-peak source levels. In fact, the highest zero-to-peak pressure levels from man-made sources probably were created by the nuclear tests conducted in the ocean, in the atmosphere over the ocean, and on oceanic islands in the past half century. In the case of chemical explosive devices, they are physically small, portable, and easily deployed from a variety of platforms. They have traditionally been used for research purposes and some years ago were incorporated into a naval antisubmarine warfare sensor system. Explosives are also used in construction and removal of unwanted undersea structures. In the past, they were used as sources in seismic exploration, but modern-day surveys employ air-gun arrays. However, a few geophysical research programs continue to use explosives.

Military vessels undergo a series of tests to determine their ability to withstand explosions near the hull. Trials, known as ship shock trials, are carried out on every class of U.S. military vessel hull prior to commissioning. During a ship shock trial, explosives are detonated near the hull and hull stress is measured. While the pressure waves generated by this source are very large, this noise source is extremely episodic, since few ship shock trials are conducted annually. Large explosives are also used occasionally for the "sinkex" (a ship hulk sunk with a torpedo warshot) weapon tests during development and for test firing of operational stores for military readiness exercises.

A significant literature base has resulted from research into the spectral and amplitude characteristics of chemical explosives (Figure 2-6). The source level and spectral structure, which is relatively flat, of an explosive device can be predicted using the charge weight and depth of detonation (Table 2-5). For example, the zero-to-peak source pressure level, $SL(0-pk)$, of the initial shock wave, the largest-amplitude component in the detonation time series created by a high explosive, is given by the formula

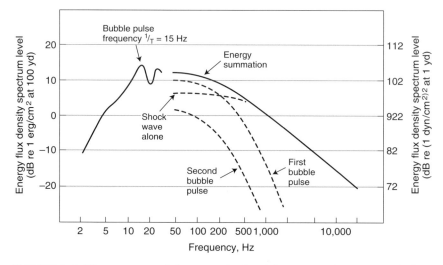

FIGURE 2-6 The spectrum of the acoustic signal from an underwater explosion. The quantity labeled "energy flux density" actually is the instantaneous pressure amplitude squared summed over the duration of the signal, as discussed in the Glossary. To convert the units on the right-hand vertical axis from dB re 1 [(dynes/cm^2)^2](sec) at 1 yd into dB re 1 (μPa^2)(sec) at 1 m, approximately 100 should be added to the values so that the resulting axis extends from 162 to 222 dB re 1 (μPa^2)(sec) at 1 m. These spectral levels pertain to a 1-lb. charge detonated at a depth of 20 fathoms (36.6 m) and are equal to the actual source level at each frequency for a signal of 1-sec duration. The corresponding broadband zero-peak pressure level at 1 m from the source for the initial shock wave from a 1-lb. charge of TNT is 272 dB re 1 μPa at 1 m, as given by Equation 2-1 in the text. The plot shows the addition of the shock-wave and bubble pulse energies at frequencies greater than 1/T, with T equal to the time interval between the shock wave and the first bubble pulse.
SOURCE: Urick, 1975. Reprinted with permission from the Acoustical Society of America.

$$SL(0\text{--}pk)(dB \text{ re } 1 \ \mu Pa \text{ at } 1 \text{ m}) = 271.8 \text{ dB} + 7.533*\log(w) \qquad (2\text{-}1)$$

where w is the charge weight in pounds. The third column of Table 2-5 provides some examples from the use of Equation 2-1.

Industrial and Construction Contributions to Marine Noise

The range of activities in this category is extremely broad, ranging from power plants located near the seaside to pile driving, dredging, shipyards, canal lock structure operations, and general harbor daily functions. The coupling of this energy, which is a combination of terrestrially based to shoreline and nearshore, into the marine environment is poorly understood.

TABLE 2-5 Zero-to-Peak Pressure Level and Spectral Level at 1 kHz of
Pressure Amp Squared Times Duration for High Explosive Detonated at
40 m Depth

TNT (lb.)	Spectral Level at 1 kHz of Pressure Amp Squared Time Duration (dB re 1 μPa at 1 m)	Zero-to-Peak Pressure Level at 1 m (dB re 1 μPa²s at 1 m)
1	192	272
10	200	279
100	207	287

SOURCE: Urick, 1975. Courtesy of McGraw-Hill.

This broad range of activities produces a range of source levels and acoustic
patterns:

- pile driving (impulsive, very high amplitudes),
- power plants (very strong 60-cycle and harmonics),
- industrial (tones at frequencies of machinery operating speeds),
- dredging [both shipborne machinery and mechanical motion (suction and earth-moving devices, possible explosive use)], and
- power-generating windmills.

A typical spectral structure is broadband with the superposition of a number of lines originating from reciprocating machinery or engines.

Some measurements of the underwater sounds created by these types of
sources have been presented in the open literature (e.g., Richardson et al.,
1995). Additional measurements are contained in various technical reports
and memoranda. It would be useful to gather these measurements together
into one easily accessible place (as the committee recommends) for use by
the scientific and regulatory communities as well as others. Note, however,
that the coupling of land-based vibrations and very nearshore sounds into
the offshore underwater acoustic field is highly dependent on the geology,
morphology, and length of the land-based portion of the propagation path.
Therefore, measurements made in one offshore area are not necessarily
applicable to other offshore areas. Numerical modeling of the coupling
between land-based vibrations and the ocean acoustic field is a subject of
current research. This uncertainty associated with the coupling process
makes an assessment of the overall impact of these sounds on the marine
environment difficult. However, of greater importance is to understand the
potential impact of a given noise source in its actual geological setting on
the marine ecosystems that are located nearby. At present, the evaluation
for land-based and very nearshore sources probably is best done using
actual underwater acoustic measurements in the region of interest.

LONG-TERM TRENDS IN OCEAN NOISE

One of the most important and challenging issues that emerges in an examination of ocean noise and its effects on marine life is the quest to determine any long-term trends in the overall levels of sound in the sea. How has noise in the sea increased with time since the 1850s through increased industrialization and related maritime activities? What parameters, other than direct noise measurements, might be related to the overall sound levels produced by the myriad of sources described? What, if any, modeling capabilities exist to predict ocean noise levels and other noise characteristics in the decades to come? Is there any hope that humans might influence these predicted changes through time by introducing appropriate mitigation measures? Answering these questions holds enormous significance for life in the sea; after all, long-term changes in background noise levels may influence animal behavior and impact the very existence of a particular species. Long-term trends are particularly insidious in that they result from the gradual accumulation of effects over much shorter periods of time for which these effects may appear to be imperceptible.

Although the importance of assessing long-term trends in ocean noise levels is clear, there is a remarkable dearth of theories or data addressing this topic. Commercial shipping noise is apparently the only area in which it is possible to make informed comments concerning long-term trends, and even in that case, the data sets are very limited and the discussion is usually speculative. The focus on shipping implies an emphasis on frequencies of a few hundred hertz and below and a geographical bias toward the northern hemisphere, where most of the dominant shipping lanes exist. In this section, the first attempts to estimate the preindustrial noise background by examining measurements in areas of the South Pacific with extraordinarily low ship traffic are described (Cato, 1997, 2001). The addition of the anthropogenic component of noise during the Industrial Revolution, principally the result of shipping, is reviewed, followed by a summary of existing data on long-term trends in shipping noise levels and a discussion of various indicators for evaluating and predicting shipping noise levels. Finally, speculations on the long-term trends in ocean sounds other than those from shipping are presented. Recommendations for future research to measure and predict long-term trends in ocean noise are listed in Chapter 5.

Preindustrial Noise

One approach to modeling long-term trends is to hypothesize that the overall background noise level remained essentially constant until the onset of the Industrial Revolution in 1850. At that time, land-based industrial activities began to escalate rapidly and resulted in an enormous increase in the use of ships under power to transport goods and provide services over the oceans. Other related, though less significant, developments were the powering of the world's naval vessels and the expansion of coastal and offshore construction activities. Additional, though also secondary, sources of anthropogenic noise that emerged much later (primarily during the past 50 years) were those produced by offshore oil exploration and drilling activities, naval sonars, and acoustical oceanographic research. This model is based on the assumption that the noise contributed by natural physically generated and biological sources is independent of time and that human-generated noise prior to the Industrial Revolution was negligible. In fact, this assumption is open to debate, since there is some evidence that global climate change effects have resulted in higher sea states (Bacon and Carter, 1993; Graham and Diaz, 2001); these could potentially cause an increase in the noise levels generated by breaking waves over time. This effect is very likely of secondary importance, however, and therefore the model in which commercial shipping provides the primary time-dependent influence on long-term noise levels is adopted here.

The waters surrounding Australia provide a unique opportunity to estimate preindustrial noise levels. Shipping densities in some areas are extremely low or negligible. This situation provided Cato (1997b, 2001) with the opportunity to determine the "usual lowest ocean noise" level from an extensive suite of measurements (Plate 5). There are several striking features that appear in these data. First, the lowest noise level decreases monotonically from 55 dB re 1 μPa^2 per Hz at 10 Hz to 30 dB re 1 μPa^2/Hz at 11 kHz. Second, the ship traffic noise data indicate values as high as 80 dB re 1 μPa^2/Hz at 20 Hz in the Tasman Sea, "and these approach the traffic noise levels of North American and European waters" (Cato, 2001), consistent with the Wenz curves (cf. Plate 1). In fact, at 20 Hz the ship noise level exceeds Cato's lowest level by about 25 dB. On the other hand, in the Timor and Arafura seas, the ship noise levels vary from 50 to 58 dB re 1 μPa^2 per Hz in the band 20-200 Hz and exceed the lowest level by only a few decibels. Third, Cato (2001) points out that the level of naturally generated noise, both physical and biological, frequently equals or exceeds that produced by ship traffic at 200 Hz and below. Specifically, the Australian data indicate that the wind wave noise continues to increase below 200 Hz, in contrast to the behavior of the deep-water Wenz curves[5] for various sea states, which are based on northern hemisphere data. Cato suggests that this is probably due to the difficulty of separating the effects of ships and breaking waves in northern waters.

---

[5]Shallow-water measurements, including those of Wenz (1962), Piggott (1964), and Arase and Arase (1967), indicate an increase in noise levels produced by wind waves at frequencies below 100 Hz.

Postindustrial Noise

With the Cato data providing a glimpse at the preindustrial noise background, it is now appropriate to examine any existing data on long-term trends in noise levels. As mentioned earlier, the paucity of data on this topic is surprising. However, it is encouraging that the few existing data sets are consistent with one another.

First, consider the data and interpretation provided by Ross (1976, 1993). Long-term trends in ambient noise levels can be observed at low frequencies (unspecified, but presumably below 200 Hz) in the East Pacific and East and West Atlantic Oceans (Figure 2-7). Ross (1993) concluded from these data that "low-frequency noise levels increased by more than 10 dB in many parts of the world between 1950 and 1975," corresponding to about 0.55 dB per year. This increase was attributed to two factors associated with commercial shipping, namely a doubling of the number of ships, which accounts for an increase of 3-5 dB, and greater average ship speed, propulsion power, and propeller tip speed, which are responsible for at least an additional 6 dB. Ross (1976) also speculated that, during the next quarter century, ship noise levels may increase by only about 5 dB because "the number of ships may be expected to increase only about 50 percent and the noise per ship by only a few dB."

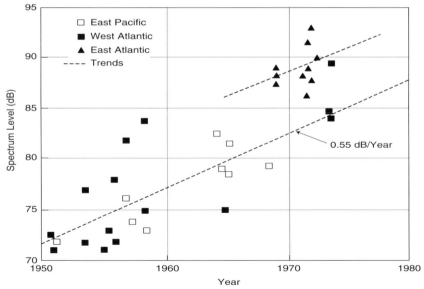

FIGURE 2-7 Long-term trend for low-frequency ambient levels for period 1958–1975. SOURCE: Ross, 1993, courtesy of Acoustics Bulletin.

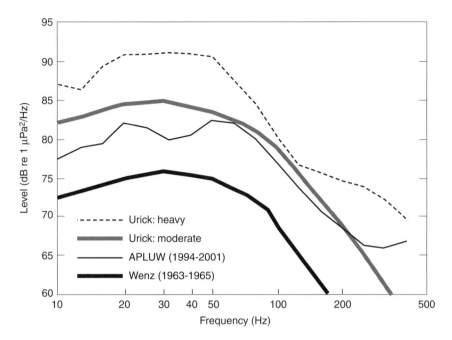

FIGURE 2-8 Point Sur autospectra compared with Wenz (1969). Point Sur data are converted to one-third octave levels and then normalized by the third-octave bandwidths for direct comparison. Shown for reference are the "heavy" and "moderate" shipping average deep-water curves presented by Urick. SOURCE: Andrew et al., 2002. Reprinted with permission of the Acoustical Society of America.

In another attempt to assess long-term trends, Andrew et al. (2002) compared noise measurements made on a receiver on the continental slope of Point Sur, California, from 1994 to 2001 with those collected on the same receiver from 1963 to 1965 (Wenz, 1969). The results of their analysis indicated an increase of approximately 10 dB over 33 years (about 0.3 dB per year) from 20 to 80 Hz (Figure 2-8). Andrew et al. attributed this change principally to increases in the number and gross tonnage of commercial ships, a conclusion consistent with Ross's results. They indicated that they do not have a satisfactory explanation for the increased noise from 100 to 400 Hz (up to 9 dB) or the minimum increase of 3 dB close to 100 Hz.

Mazzuca (2001) synthesized the results of Ross (1976) and Wenz (1969) to obtain an overall 16-dB increase in low-frequency noise level from shipping during the period 1950-2000. This value corresponds to a rate of increase of 0.32 dB per year, or about 3 dB per decade.

Indicators for Evaluating and Predicting Shipping Noise

How does one quantitatively correlate ocean noise levels with shipping activity and its origins in industrial activities? Efforts to determine the principal sources of noise on ships have constituted an active area of research for quite some time. The classical model, put forward by Ross (1976), states that ship-radiated noise is directly related to ship length and speed. Yet this theory was criticized recently by Wales and Heitmeyer (2002), who contend that there is a "negligible correlation between the source level and the ship speed and the source level and the ship length." Part of the reason for this lack in correlation may be due to the type of source model used by Wales and Heitmeyer (Heitmeyer, personal communication, 2002). In any case, these observations complicate attempts to determine the principal ship parameters affecting the overall noise levels associated with a large number of ships. Nevertheless, it is possible to make some general well-founded comments regarding ship-radiated noise and shipping traffic and their possible implications for long-term ocean noise levels.

First, there is no doubt that ships generate noise, principally by propeller cavitation and machinery. Second, it is well known that aging ships tend to generate more noise as mechanical and electrical systems deteriorate over time. Third, newer ships have a number of noise-mitigating characteristics, including quieter diesel-electric propulsion systems and deeper propellers that are less prone to cavitation. Fourth, and most important, the number of ships and gross tonnage of the world fleet have increased substantially since 1950 (Figure 2-9) (McCarthy and Miller, 2002). During this period, the number of ships almost tripled (from 30,000 to 87,000 ships), while the gross tonnage increased by a factor of about 6.5 (from 85 to 550 million gross tons). Interestingly, the logarithmic (dB) equivalent of a factor of 6.5 is about 16 dB, exactly corresponding to the observed increase in low-frequency noise levels. These data suggest the following simple relationship between changes in noise levels and gross tonnage:

$$\text{Change in shipping noise (dB)} = 20\log_{10}\left[\frac{\text{final gross tonnage}}{\text{initial gross tonnage}}\right] \quad (2\text{-}2)$$

While Equation 2-2 is highly speculative and its predictive capability must be tested (other parameters, such as changes in ship speed, may need to be included, although they may all be correlated with gross tonnage), there is no doubt that world economic conditions influence shipping activity, which in turn affects overall noise levels in the sea. Westwood et al. (2002) estimate "that over 90 percent of world trade is carried by sea and over the period 1985 to 1999, world seaborne trade increased by 50 percent to

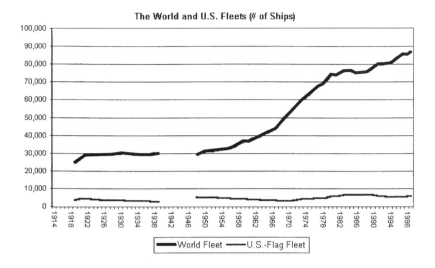

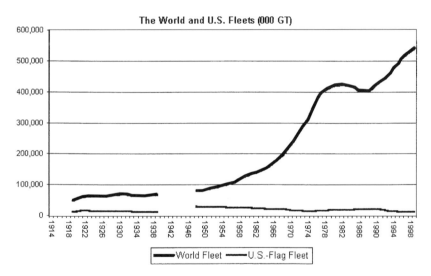

FIGURE 2-9 Global shipping fleet trends, 1914–1998. Only those commercial ships registered in the U.S. (the U.S. flag fleet) are subject to U.S. regulations and laws when operating outside U.S. territorial waters. SOURCE: McCarthy, 2001. Courtesy of http://coultoncompany.com.

about 5 billion tons with the largest increase coming in crude oil and oil products shipments. During 1990-1998 growth averaged 3.2 percent per annum." The 50 percent trade increase is comparable to the 38 percent increase in gross tonnage during the same period. Applying Equation 2-2 gives a result, $20*\log(1.38) = 2.8$, which is not much different from the expected increase in noise levels of 4.5 dB (equal to 0.32 dB/year multiplied by 14 years) for this same time period. Only further study can elucidate whether the similarity in these figures is purely coincidental or scientifically meaningful.

Long-Term Trends in Other Sources of Ocean Noise

No long-term systematic ocean acoustics data set exists to permit a scientific assessment of trends of noise in the ocean. Therefore, the following discussion speculates on possible trends rather than describing any. Although the levels of naturally occurring sound from physical sources (particularly wind-generated and ice-generated noise) may be changing as a result of possible changes in weather patterns associated with global warming, these changes are believed to be dwarfed by other trends. Ocean measurements do exist that demonstrate that ambient sound from some biological sources is increasing in a few locations in the world, for example, sounds produced by humpback whales in the waters around Australia (Cato and McCauley, 2002). The noise associated with whales is expected to asymptote at preexploitation levels as the whale populations return to their preexploitation numbers. However, the overall trend in noise from all biological sources is unknown.

Regarding anthropogenic noise sources, the previous sections of this chapter show that educated speculation (Ross, 1976; Mazzuca, 2001) and measurements at one location (Andrew et al., 2002) suggest that shipping noise at low frequencies (20-80 Hz) has increased by about 10-15 dB over a 25-50-year period. Although the decrease in detection range associated with this increase in noise can be calculated from a sonar systems perspective, the degree to which this change has an adverse impact on the marine environment is unknown. The change in level itself is not a cause of great concern given that naturally occurring processes can change noise levels by 20-30 dB over short periods (e.g., Plate 1). However, other properties of this increase in shipping noise may be biologically important, such as the increase in the prevalence of noise (decrease in time intervals between shipping-noise-dominated periods or increase in the number of locations where shipping noise is a significant contributor), the character of the shipping-generated signals themselves, and so on. Increases in the number and size of commercial and recreational craft have resulted in noise-level increases substantially greater than 10 dB in some areas (e.g., 30 dB or so in the frequency band from 10 to 100 Hz in Singapore Harbor) (Potter and Delory, 1998), but the potential impact on these environments is unknown. This trend has been accompanied by the proliferation of boats and ships equipped with depth sounders and fish finders, which have likely raised the high-frequency (above a few kilohertz) noise in some localized areas. Again, the amount of increase and its potential effects are unknown.

Trends in seismic exploration are much simpler to define in terms of activity than in terms of contribution to the underwater sound field. As discussed previously in this chapter, industry publications periodically report the numbers of surveys presently being conducted in general locations. However, given that exploration methods have been changing, for example, large explosive sources have been replaced by air-guns, which have evolved into air-gun arrays that focus the radiated acoustic energy in the vertical direction, and that undiscovered oil and gas reserves probably are deeper within the earth and/or are to be found in deeper waters, the overall impact of changes on the ocean sound field is difficult to evaluate without a combined ocean noise measurement and numerical modeling effort.

In contrast to focusing of acoustic energy in the vertical by present-day geophysical exploration sources, the newly developing low-frequency navy sonars radiate acoustic energy preferentially in the horizontal direction. Because of the very low absorption of sound in the ocean at low frequencies, these active sonar signals can travel over large distances. Another recent trend in U.S. military sonar has been toward the use of active systems in coastal and shallow-water regions. These sonars have the potential to adversely impact marine mammals; evidence indicates that navy mid-frequency (1-10 kHz) tactical sonars were directly related to the March 2000 mass stranding of marine mammals in the Bahamas (Evans and England, 2001). Other mass strandings have been associated with the transmission of high-level sonar signals, for instance, the May 1996 event in the Mediterranean Sea (D'Amico and Verboom, 1998). Whatever the frequency band, the growth of sonar activity for military purposes started from essentially zero to the present-day levels over just the last half century or so.

The number and type of man-made explosions that affect the world's oceans also have been changing. The largest of these events is associated with nuclear tests, which have taken place only since 1945 (Lawson, 2002). Many of the early tests by the United States from 1945 to 1962 were atmospheric tests conducted on small islands in the central part of the Pacific Ocean just north of the equator. The energy released during these tests certainly created high-level impulsive ocean acoustic signals that traveled over great distances. Much of the subsequent U.S. testing was done underground, sufficiently far from the coast that little impact on the ocean environment occurred. However, underground tests near coastlines can create high levels of underwater sound. As an illustration, one of the French nuclear tests in 1995-1996 on the Mururoa Atoll in French Polynesia in the South Pacific Ocean generated underwater signals that were recorded by a single omnidirectional hydrophone at a range of 6,670 km with levels 20-45 dB above background noise across the frequency band 2-30 Hz (D'Spain et al., 2001). Given the steady progress since the mid-1990s in the number of nations that have signed, and have ratified, the Comprehensive Nuclear Test Ban Treaty, these tests appear to be increasingly rare.

Long-term trends in the use of chemical explosive devices also may be taking place. Any speculation on these trends must exclude times of war (the underwater noise created by explosions during the great naval battles in World War II must have been extremely high) since the occurrence of war and resulting contributions to the ocean noise field are highly unpredictable and extremely episodic. Long-term trends in the use of smaller explosive devices also may be taking place. As mentioned previously, explosive sources used in seismic exploration are being replaced by air-gun arrays. However, explosives are routinely used to sever abandoned wellheads so that they can be removed and to decommission the rigs themselves. As oil production in a given region matures and declines, the use of explosives in this way increases. The use of explosives in ocean acoustic and geophysical research has decreased, but these sources still are deployed in a few experiments. Military use of explosive charges as the source component in active sonar systems (e.g., SUS; Urick, 1975) appears to be decreasing. Hull shock tests are rare events and do not appear to be changing significantly in frequency of occurrence. The use of seal bombs has been discouraged by U.S. and international regulations and is being replaced by other types of acoustic deterrent devices. Fishing by the detonation of underwater explosives (a technique whose success improves with its increasing adverse impact on the marine environment) is banned but still is practiced in some regions. In any case, one quantitative measure of the long-term change in numbers and spatial distribution of underwater explosions is possible to obtain, at least for the North Pacific Ocean. The number and estimated source locations of detonations recorded over a modern-day period of time could be compared to those recorded by 20 Missile Impact Location System hydrophones over a one-year period from August 1965 to July 1966 (following Spiess et al., 1968). Nearly 20,000 explosions were detected within this one-year period, with the winter rate of occurrence of 300 explosions per month increasing to 4,000 explosions per month in summer. The highest activity was detected off the west coast of North America, in the Gulf of Alaska, north of Hawaii, and seaward of the Japanese and Kuril Islands (Spiess et al., 1968). The significance of any of these possible changes in the occurrence of underwater explosions to the marine environment is unknown.

3

Effects of Noise on Marine Mammals

INTRODUCTION

Richardson et al. (1995) provided a comprehensive summary of published and gray literature data on marine mammal responses to specific noise sources. Although the literature continues to expand and many valuable new studies have appeared, most recent publications have tended to provide variations on themes rather than new data at variance with the conclusions summarized by Richardson et al.. A number of factors affect the response of marine mammals to sounds in their environment: the sound level and other properties of the sound, including its novelty; physical and behavioral state of the animal; and prevailing acoustic characteristics and ecological features of the environment in which the animal encounters the sound. Critical issues about what determines effects of and responses to intense transient sounds and what are the effects of long-term anthropogenic sound on individuals and populations remain unanswered (see Box 3-1 for the priority research areas identified by the NRC [2000]). The indirect effects of anthropogenic sound on marine mammals via effects on their predators, prey, and other critical habitat elements are largely uninvestigated.

HEARING CAPABILITIES OF MARINE ORGANISMS

Marine Mammals

Hearing research has traditionally focused on mechanisms of hearing loss in humans. Animal research has therefore emphasized experimental

---

Box 3-1
Priority Research for Whales and Seals
Recommended by NRC (2000)

To move beyond requiring extensive study of each sound source and each area in which it may be operated, [NRC (2000) recommended that] a coordinated plan should be developed to explore how sound characteristics affect the responses of a representative set of marine mammal species in several biological contexts (e.g., feeding, migrating, and breeding). Research should be focused on studies of representative species using standard signal types, measuring a standard set of biological parameters, based on hearing type (Ketten, 1994), taxonomic group, and behavioral ecology (at least one species per group). This could allow the development of mathematical models that predict the levels and types of noise that pose a risk of injury to marine mammals. Such models could be used to predict in multidimensional space where temporary threshold shift (TTS) is likely (a "TTS potential region") as a threshold of potential risk and to determine measures of behavioral disruption for different species groups. Observations should include both trained and wild animals. The results of such research could provide the necessary background for future environmental impact statements, regulations, and permitting processes.

Groupings of Species Estimated to Have Similar Sensitivity to Sound
Research and observations should be conducted on at least one species in each of the following seven groups:

1. Sperm whales (not to include other physterids)
2. Baleen whales
3. Beaked whales
4. Pygmy and dwarf sperm whales and porpoises [high-frequency (greater than 100 kHz) narrowband sonar signals]
5. Delphinids (dolphins, white whales, narwhales, killer whales)

---

work on ears in other species as human analogs. Consequently, researchers have generally investigated either very basic mechanisms of hearing or induced and explored human auditory system diseases and hearing failures through these test species. Ironically, because of this emphasis, remarkably little is known about natural, habitat, and species-specific aspects of hearing in most mammals. Marine mammals represent an extreme example of not only habitat adaptations but also adaptations in ear structure and hearing capabilities.

The same reasons that make marine mammals acoustically and auditorally interesting—that is, that they are a functionally exceptional *and* an aquatic ear—also make them difficult research subjects. Some issues about marine mammal hearing can be addressed both directly and inferentially from the data at hand. While large gaps remain in our knowledge, progress has been made on some fronts related to sound and potential impacts from noise.

6. Phocids (true seals) and walruses
7. Otarids (eared seals and sea lions)

Signal Type
Standardized analytic signals should be developed for testing with individuals of the preceding seven species groups. These signals should emulate the signals used for human activities in the ocean, including impulse and continuous sources.

1. Impulse—air-guns, explosions, sparkers, sonar pings
2. Continuous—frequency-modulated [low-frequency (LFA) and other sonars], amplitude-modulated (drilling rigs, animal sounds, ship engines), broadband (ship noise, sonar)

Biological Parameters to Measure
When testing representative species, several different biological parameters should be measured as a basis for future regulations and individual permitting decisions. These parameters include the following:

- Mortality
- TTS at signal frequency and other frequencies
- Injury—permanent threshold shifts
- Level B harassment
- Avoidance
- Masking (temporal and spectral)
- Absolute sensitivity
- Temporal integration function
- Nonauditory biological effects
- Biologically significant behaviors with the potential to change demographic parameters such as mortality and reproduction.

SOURCE: NRC, 2000.

Marine mammals, and whales in particular, present an interesting hearing paradox. On one hand, marine mammal inner ears physically resemble land mammal inner ears, although the external ears are typically absent and the middle ear extensively modified. Since many forms of hearing loss are based in physical structure of the inner ear, it is likely hearing damage occurs by similar mechanisms in both land and marine mammal ears. On the other hand, the sea is not, nor was it ever, even primordially silent. Whales and dolphins, in particular, evolved ears that function well within this context of natural ambient noise. This may mean they developed "tough" inner ears that are less subject to hearing loss under natural ocean noise conditions. Recent anatomical and behavioral studies do indeed suggest that whales and dolphins may be more resistant than many land mammals to temporary threshold shifts (TTSs), but the data show also that they are subject to disease and aging processes. This means they are not immune to hearing loss, and certainly, increasing ambient noise via human activities is a reasonable candidate for exacerbating or accelerating such losses.

Unfortunately, existing data are insufficient to predict accurately any but the grossest acoustic impacts on marine mammals. Little information exists to describe how marine mammals respond physically and behaviorally to intense sounds and to long-term increases in ambient noise levels.

The data available show that all marine mammals have a fundamentally mammalian ear, which through adaptation to the marine environment has developed broader hearing ranges (Figure 1-1) than are common to land mammals. Audiograms are available for only 10 species of odontocetes and 11 species of pinnipeds. All are smaller species that were tested as captive animals. However, there are 119 marine mammal species, and the majority are large, wide-ranging animals that are not approachable or testable by normal audiometric methods. Therefore, direct behavioral or physiologic hearing data for nearly 80 percent of the genera and species of concern for coastal and open-ocean sound impacts do not exist. For those species for which no direct measure or audiograms are available, hearing ranges are estimated with mathematical models based on ear anatomy obtained from stranded animals or inferred from emitted sounds and controlled acoustic exposure experiments in the wild.

The combined data from audiograms and models show there is considerable variation among marine mammals in both absolute hearing range and sensitivity. Their composite range is from ultra- to infrasonic. Odontocetes, like bats, are excellent echolocators, capable of producing, perceiving, and analyzing ultrasonic frequencies well above any human hearing. Odontocetes commonly have good functional hearing between 200 and 100,000 Hz, although some species may have functional ultrasonic hearing to nearly 200 kHz. The majority of odontocetes have peak sensitivities (best hearing) in the ultrasonic ranges, although most have moderate sensitivity to sounds from 1 to 20 kHz. No odontocete has been shown audiometrically to have acute, that is, best sensitivity or exceptionally responsive, hearing (<80 dB re 1 µPa) below 500 Hz.

Based on functional models, good lower-frequency hearing appears to be confined to larger species in both the cetaceans and pinnipeds. No mysticete has been directly tested for any hearing ability, but functional models indicate their hearing commonly extends to 20 Hz, with several species, including blue, fin, and bowhead whales, that are predicted to hear at infrasonic frequencies as low as 10–15 Hz. The upper functional range for most mysticetes has been predicted to extend to 20–30 kHz.

Most pinniped species have peak sensitivities between 1 and 20 kHz. Some species, like the harbor seal, have best sensitivities over 10 kHz. Only the northern elephant seal has been shown to have good to moderate hearing below 1 kHz (Kastak and Schusterman, 1999). Some pinniped species are considered to be effectively double-eared in that they hear moderately well in two domains, air and water, but are not particularly acute in either. Others, however, are clearly best adapted for underwater hearing alone.

To summarize, marine mammals as a group have functional hearing ranges of 10 Hz to 200 kHz. They can be divided into infrasonic balaenids (probable functional ranges of 15 Hz to 20 kHz; good sensitivity from 20 Hz to 2 kHz); sonic to high-frequency species (100 Hz to 100 kHz; widely variable peak spectra), and ultrasonic dominant species (200 Hz to 200 kHz general sensitivity; peak spectra 16-120 kHz) (Wartzok and Ketten, 1999).

Other Marine Organisms

The inner ear of fishes and elasmobranchs (sharks and rays) is very similar to that of terrestrial vertebrates [see Popper and Fay (1999) for review]. While there are data on hearing capabilities for fewer than 100 of the 25,000 extant species, investigations of the auditory system of evolutionarily diverse species support the suggestion that hearing is widespread among virtually all fishes, as well as elasmobranchs.

Most species of fish and elasmobranchs are able to detect sounds from well below 50 Hz (some as low as 10 or 15 Hz) to upward of 500-1,000 Hz (Figure 3-1).[1] Moreover, a number of fish species have adaptations in their auditory systems that enhance sound detection and enable them to detect sounds to 3 kHz and above and have better sensitivity than nonspecialist species at lower frequencies. Goldfish and American shad are examples of specialist species, while Atlantic salmon and Atlantic cod are examples of species without specializations.

There are very few data on hearing by marine invertebrates, although a number of species have highly sophisticated structures, called statocysts, that have some resemblance to the ears of fishes (Offutt, 1970; Budelmann, 1988, 1992). The statocysts found in the cephalopods (octopods and squid) may primarily serve for determination of head position in a manner similar to the components of the vertebrate ear that determine head position for vestibular senses. It is possible, but not yet demonstrated, that cephalopods use their statocysts for detection of low-frequency sounds.

There is also some evidence that a number of crustacean species, such as crabs, have statocysts that are somewhat similar to those found in cephalopods, although they have evolved separately. While there are no data for

---

[1]It is also important to note that there are far fewer data for sharks than for bony fishes, and the studies were usually based on one or two animals. Thus, all shark data must be taken as somewhat tentative. Since sharks make up such an important part of the marine ecosystem, and since sharks rely heavily on sound to detect prey, it would be of great value to have additional data on hearing in at least several species.

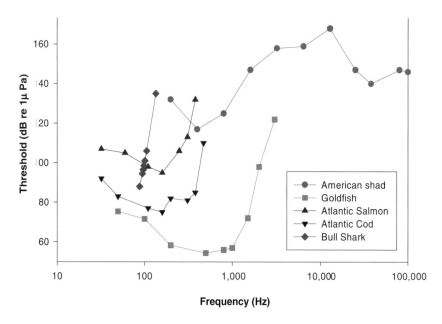

FIGURE 3-1 Fish and shark audiograms. Hearing capabilities in several fish species and a shark showing the lowest sound level that an animal can detect at each frequency. It is important to note that while thresholds here are presented in units of pressure, it is very likely that a number of species, including the sharks, respond best to particle acceleration and had experiments been done in terms of acceleration the shapes of the hearing curves might be somewhat different, though it is likely that the range of detection would not change very much. The stimuli in some of these experiments were in the near field where particle acceleration and pressure are not directly related. SOURCES: American shad: Mann et al. (1997); goldfish: Jacobs and Tavolga (1967); Atlantic salmon: Hawkins and Johnstone (1978); Atlantic cod: Chapman and Hawkins (1973); bull shark: Kritzler and Wood (1961).

hearing by marine crabs, a number of species of semiterrestrial fiddler and ghost crabs are not only able to detect sounds but also use special sounds for communication (reviewed in Popper et al., 2001). In addition, a number of physiological studies of statocysts of marine crabs suggest that some of these species are potentially capable of sound detection (Popper et al., 2001).

Marine reptiles include snakes and turtles. Although marine snakes have auditory systems similar to those of terrestrial snakes, nothing is known about their acoustic abilities. Despite considerable interest in marine turtles, since many species are endangered, very little is known about their hearing. Difficulties in developing methods to successfully train turtles to respond to acoustic stimuli have hindered research in this area. Ears of turtles are well developed, and there is some evidence that at least a few species of marine turtles can detect sounds below 1 kHz. However, until more data are available, this value must be taken with considerable caution. Bartol et al. (1999) measured the hearing of 35 juvenile loggerhead sea turtles and the results suggested a hearing range from at least 250-750 Hz, with the most sensitive threshold recorded at the lowest frequency tested, 250 Hz. Ridgway et al. (1969) found that green turtles were most sensitive to frequencies between 300 and 400 Hz and sensitivity declined rapidly at frequencies outside of this range. There is some additional evidence from attempts at behavioral studies and from recordings of responses of the inner ear, but no data suggest higher frequencies of hearing.

ACOUSTIC TRAUMA IN MARINE MAMMALS

Recent reports and retrospectively analyzed data show an association between the use of multiple high-energy mid-range sonars and mass strandings of beaked whales (*Ziphius cavirostris*). Recent mass strandings of beaked whales have occurred in a temporal and spatial association with ongoing military exercises employing multiple high-energy, mid-frequency (1-10 kHz) sonars. Strandings in the Mediterranean (D'Amico and Verboom, 1998), the New Providence Channel in the Bahamas (Evans and England, 2001), and most recently in the Canary Islands (2002) have greatly increase public awareness of the issue of noise in the ocean. In addition, a retrospective review of earlier beaked whale strandings suggests that there is at least an indirect causal relationship between the strandings and the use of multiple, mid-range sonars in military exercises in some nearshore areas. Although the correlation in time between the use of sonars and the strandings is quite compelling, there is no clear demonstration as yet of any causal mechanism. Acoustic trauma is a very explicit form of injury. In the beaked whale cases to date, the traumas that were observed could result from many causes, both directly and indirectly associated with sound, or could have been from other causes. Indeed, similar traumas have been observed in terrestrial mammals under circumstances having no relation to sound exposure. Careful sampling has rarely been possible in beaked whale cases, which has made adequate diagnosis problematic. To date, only six specimens of beaked whale have been rigorously analyzed. The NATO report (D'Amico, 1998) and the joint NOAA-Navy interim report (Evans and England, 2001) have not been discussed in detail by this committee because of the preliminary nature of the findings. However, this is clearly a subject to which much additional research needs to be directed. A program should be instituted to investigate carefully the causal mechanisms that may explain the traumas observed and how the acoustics of high-energy, mid-range sonars directly or indirectly are related to them and to mass stranding events.

EFFECTS OF MARINE NOISE ON MAMMAL BEHAVIOR

Behavioral responses of marine mammals to noise are highly variable and dependent on a suite of internal and external factors. Internal factors include

- individual hearing sensitivity, activity pattern, and motivational and behavioral state at time of exposure;
- past exposure of the animal to the noise, which may have led to habituation or sensitization;
- individual noise tolerance; and
- demographic factors such as age, sex, and presence of dependent offspring.

External factors include

- nonacoustic characteristics of the sound source, such as whether it is stationary or moving;
- environmental factors that influence sound transmission;
- habitat characteristics, such as being in a confined location; and
- location, such as proximity to a shoreline.

Behavioral responses range from subtle changes in surfacing and breathing patterns, to cessation of vocalizations, to active avoidance or escape from the region of the highest sound levels.

Typical changes in cetacean response to anthropogenic noise are summarized from several studies of bowhead whales as shorter surfacings, shorter dives, fewer blows per surfacing, and longer intervals between successive blows (Richardson et al., 1995). These subtle changes are often the only observable reaction of whales to reception of anthropogenic stimuli. Although there may be statistically significant changes in some of these subtle behavioral measures, there is no evidence that these changes are biologically significant for the animals. Typical changes in vocalizations are a reduction or cessation in calling as shown in right whales in response to boats (Watkins, 1986); bowhead whales in response to playbacks of industrial sounds (Wartzok et al., 1989); sperm whales in response to short sequences of pulses from acoustic pingers (Watkins and Schevill, 1975); and sperm and pilot whales (*Globicephala melaena*) in response to the Heard Island Feasibility Test source (Bowles et al., 1994). Humpback whales, which appeared in all other behavioral measures to have habituated to the presence of whale-watching boats, still tended to cease vocalizations when near boats (Watkins, 1986).

Not all cetaceans respond with a decrease or cessation of calls. Sperm whales continued calling when encountering continuous pulsing from echo sounders (Watkins, 1977) and when exposed to received sound levels of 180 dB re 1 μPa (RMS) from the discharge of a detonator (Madsen and Møhl, 2000); humpback whales moved away from low-frequency (3-kHz range) sonar pulses and sweeps but did not change their calling (Maybaum, 1993); and a fin whale continued to call with no change in rate, level, or frequency components as a container ship went from idle to full power within a kilometer of the whale (Edds, 1988). Sperm whales in the Caribbean became silent in the presence of military sonar signals (3-8-kHz range; Watkins et al., 1985).

In addition to changing the frequency of occurrence of calls in the presence of noise, some species change the source level and output frequency and duration. Beluga whales adjust their echolocation clicks to higher frequencies and to higher source levels in the presence of background noise (Au et al., 1985). Miller et al. (2000) found that humpback whales exposed to low-frequency active (LFA) sonar signals increased the duration of their songs by 29 percent on average, but with a great deal of individual variation.

Given the range of observed reactions in a variety of species, it is likely that a sound that elicits escape behavior on the part of a mother and calf pair could be ignored by feeding juveniles, or actively explored by a reproductively active male. Within a given age and sex class, the cumulative probability of response by the animals is usually assumed to have a sigmoid shape with respect to increasing noise levels. Few studies have actually determined the proportion of animals responding at varying levels of acoustic signal. One study that investigated the probability of response showed that for gray whales (*Eschrichtus robustus*) the ranges broadside to a seismic gun for 10, 50, and 90 percent probability of avoidance were 3.6, 2.5, and 1.2 km, at which the received sound levels were 164, 170, and 180 dB re 1 μPa, respectively (Malme et al., 1984).

Hearing Sensitivity

Animals will only respond directly to sounds they can detect. The hearing sensitivities of only a few individuals in a select number of species are known. Even less is known about signal detection in the presence of ambient noise. Beluga whales (*Delphinapterus leucas*) can detect echolocation return signals when they are 1 dB above ambient noise levels (Turl et al., 1987), and gray whales react to playbacks of the vocalizations of a predator, the killer whale (*Orcinus orca*), when the playback signal is equal to the ambient noise (Malme et al., 1983). In both of these cases the signals have important biological significance for the animal. Anthropogenic signals do not have the same evolutionarily enhanced significance.

Many of the situationally specific responses of marine mammals to sound will be dependent on the loudness of the sound. The loudness of the sound is a function of the intensity of the sound at the location of the animal and the sensitivity of the animal to the frequencies of the sound. If the audiograms of the marine mammal species of interest are known, the potential effect of the sound can be estimated by weighting the level of the sound at each frequency by the sensitivity of the animal to that frequency, similar to the A-weighting of sound levels for humans hearing in air. Without such knowledge, it will be difficult to develop a predictive model of the impact of novel sounds on marine mammals.

Behavioral State

Animals that are resting are more likely to be disturbed by noise than are animals engaged in social activities. Würsig (personal observation cited in Richardson et al., 1995) summarized the responses of several species of dolphins to boats as "resting dolphins tend to avoid boats, foraging dolphins ignore them, and socializing dolphins may approach."

Migrating bowhead and gray whales divert around sources of noise, whether actual industrial activities or playbacks of industrial activities (Richardson et al., 1995) with almost all bowheads reacting at received levels of 114 dB re 1 µPa. However, if no other option is available, migrating bowhead whales will pass through an ensonified field to continue their migration. During spring migration, when the only available lead was within 200 m of a projector playing sounds associated with a drilling platform, the bowheads continued through a sound field with received levels of 131 dB re 1 µPa (Richardson et al., 1991).

Age and Sex

Some age and sex classes are more sensitive to noise disturbance, and such disturbance may be more detrimental to young animals. Age and sex classes can be most clearly identified and observed among pinnipeds that are on land or ice, so most of the data come from responses of these pinnipeds. Differences are expected between sexes and age among classes in the way that they respond to underwater sounds. In northern sea lions (*Eumetropias jubatus*) dominant, territory-holding males and females with young are less likely to leave a haulout site in response to an aircraft overflight than are juveniles and pregnant females (Calkins, 1979). Walrus sometimes stampede into the water in response to aircraft overflights. These stampedes sometimes result in the death of calves (Loughrey, 1959). Vessel approaches to walrus on ice can cause the herd to enter the water and in some cases leave calves stranded in slippery depressions on the ice. These calves are more vulnerable to predation by polar bears (Fay et al., 1984). Mother-calf gray whale pairs appear to be particularly sensitive to disturbance by whale-watching boats (Tilt, 1985). Humpback whale groups containing at least one calf were more responsive to approaches by small boats on several behavioral measures of respiration, diving, swimming, and aerial behaviors than were groups without a calf (Bauer et al., 1993).

Noise Source Context and Movement

The responses of cetaceans to noise sources are often dependent on the perceived motion of the sound source as well as the nature of the sound itself. For a given source level, fin and right whales are more likely to tolerate a stationary source than they are one that is approaching them (Watkins, 1986). Humpback whales are more likely to respond at lower received levels to a stimulus with a sudden onset than to one that is continuously present (Malme et al., 1985). These startle responses are one reason many seismic surveys are required to "ramp up" the signal so fewer animals will experience the startle reaction and so that animals can vacate the area of loudest signals. There is no evidence, however, that this action reduces the disturbance associated with these activities. The ramp-up of a playback signal or a seismic air-gun array takes place over a short timescale (a few tens of minutes maximum) compared to the changing received levels an animal experiences as it swims toward a stationary signal source. Bowheads react to playback levels of drill ship noise at levels they apparently tolerate quite well when they swim close to operating drill ships. Richardson et al. (1995) provide two explanations for these behavioral differences. First is the speed of ramp-up, as noted earlier. Second, the whales seen near an operating drill ship may be the ones that are more tolerant of noise. The sensitive whales seen responding to the playback levels may have already avoided the actual drill ship at ranges that were undetected by observers near the ship.

Responses of animals also vary depending on where the animals are when they encounter a novel noise source. Pinnipeds generally show reduced reaction distances to ships when the animals are in the water compared to when they are hauled out. Swimming walrus move away from an approaching ship at ranges of tens of meters, whereas walrus hauled out leave the ice at ranges of hundreds of meters (Fay et al., 1984). Similar differences in avoidance ranges have been seen in California sea lions and harbor seals. Sight and smell might also be important cues for hauled-out animals.

Bowhead whales in shallow water are more responsive to the overflights of aircraft than are bowheads in deeper water (Richardson and Malme, 1993). Beluga whales are more sensitive to ship noise when they are confined to open-water leads in the ice in the spring (Burns and Seaman, 1985). Migrating gray whales diverted around a stationary sound source projecting playbacks of LFA sonar when the source was located in the migratory path but seemed to ignore the sound source when it was located seaward of the migratory path. When the source was in the path, received levels of 140 dB re 1 μPa were sufficient to cause some path deflection. However, when the source was located seaward of the migratory path, the whales ignored source levels of 200 dB re 1 μPa at 1 m and received levels greater than 140 dB re 1 μPa (Tyack and Clark, 1998).

Variability of Responses

The range of variability of responses of marine mammals to anthropogenic noise and other disturbances can be summarized in the responses of beluga whales to ships. One of the most dramatic responses in any species of marine mammal has been observed over several years in beluga whales in the Canadian high arctic during the spring. At distances of up to 50 km from icebreakers, or other ships operating in deep channels, beluga whales respond with a suite of behavioral reactions (LGL and Greeneridge, 1986; Cosens and Dueck, 1988; Finley et al., 1990). The reactions include rapid swimming away from the ship for distances up to 80 km; changes in surfacing, breathing, and diving patterns; changes in group composition; and changes in vocalizations. The initial response occurs when the higher-frequency components of the ship sounds, those to which the beluga whale are most sensitive, are just audible to the whales. Possible explanations for this unique sensitivity to ship sounds are partial confinement of whales by heavy ice, good sound propagation conditions in the arctic deep channels in the spring, and lack of prior exposure to ship noise in that year (LGL and Greeneridge, 1986). Supporting the latter point is the observation that beluga whales that fled icebreaker noise at received levels between 94 and 105 dB re 1 μPa returned in one to two days to the area where received icebreaker noise was 120 dB re 1 μPa (Finley et al., 1990).

Beluga whales in the St. Lawrence River appear more tolerant of larger vessels moving in consistent directions than they are of small boats, fast-moving boats, or two boats approaching from different directions. Older animals were more likely to react than younger ones, and beluga whales feeding or traveling were less likely to react than animals engaged in other activities, but when the feeding or traveling whales did react, they reacted more strongly (Blane and Jaakson, 1994). In contrast to the lower rate of observed reactions of these beluga whales to larger vessels, a study of the response of beluga whale vocalizations to ferries and small boats in the St. Lawrence River showed more persistent reactions to the ferries. The whales reduced calling rate from 3.4 to 10.5 calls per whale per minute to 0.0 or under 1.0 calls per whale per minute while vessels were approaching. Repetition of specific calls increased when vessels were within 1 km, and the mean frequency of vocalizations shifted from 3.6 kHz prior to noise exposure to frequencies of 5.2-8.8 kHz when vessels were close to the whales (Lesage et al., 1999).

In Alaska, beluga whale response to small boats varies depending on the location. Beluga whales feeding on salmon in a river stop feeding and move downstream in response to the noise from outboard motorboats, whereas they are less responsive to the noise from fishing boats to which they may have habituated (Stewart et al., 1982). On the other hand, in Bristol Bay beluga whales continue to feed when surrounded by fishing vessels and resist dispersal even when purposely harassed by motorboats (Fish and Vania, 1971).

Thus, depending on habitat, demography, prior experience, activity, resource availability, sound transmission characteristics, behavioral state, and ever-present individual variability, the response of beluga whales can range from the most sensitive reported for any species to ignoring of intentional harassment. Beluga whales also show the full range of types of behavioral response, including altered headings; fast swimming; changes in dive, surfacing, and respiration patterns; and changes in vocalizations.

Long-Term Responses

Almost all the studies conducted so far have looked at only short-term effects of anthropogenic noise on marine mammals. In most cases the observed responses have been over periods of minutes to hours. Even the dramatic response of beluga whales to icebreakers in the high arctic, in which the whales moved up to 80 km and were out of the area for one to two days, falls into the category of a transient response over the annual activity budget of the animals. The whales habituated and had reduced responses to subsequent icebreakers and ships in a given season.

Multiyear abandonment of a portion of the habitat because of human activity has been reported for Guerrero Negro Lagoon in Baja California, where shipping and dredging associated with an evaporative salt works project caused the whales to abandon the lagoon through most of the 1960s. When the boat traffic declined, the lagoon was reoccupied, first by single whales and subsequently by cow-calf pairs. By the early 1980s the number of cow-calf pairs using the lagoon far exceeded the number prior to the commencement of the commercial shipping (Bryant et al., 1984). Killer whales significantly reduced their use of Broughton Archipelago in British Columbia when high-amplitude acoustic harassment devices (AHDs) were installed to deter harbor seal predation at salmon farms. The AHDs operated between 1993 and 1999, and almost no whales were observed in the archipelago throughout most of this period. However, when the devices were removed in 1999, killer whales repopulated Broughton Archipelago within six months (Morton and Symonds, 2002).

Clearly there are opportunity costs associated with even the transient behavioral changes in response to noise. The movements require energy that might otherwise have been spent in acquiring food or mates or enhancing reproduction. Repetitive transient behavioral changes have the potential of causing cumulative stress. Even transient behavioral changes have the potential to separate mother-offspring pairs and lead to death of the young, although it has been difficult to confirm the death of the young. On the other hand, pups can be injured or killed when trampled by adults rapidly leaving a haulout in a transitory response to a disturbance.

MASKING OF ACOUSTIC CUES BY MARINE NOISE

One of the most pervasive and significant effects of a general increase in background noise on most vertebrates, including marine mammals, may be the reduction in an animal's ability to detect relevant sounds in the presence of other sounds—a phenomenon known as masking. Masking, which might be thought of as acoustic interference, occurs when both the signal and masking noise have similar frequencies and either overlap or occur very close to each other in time. Noise is only effective in masking a signal if it is within a certain "critical band" (CB) around the signal's frequency. Thus, the extent of an animal's CB at a signal's frequency, and the amount of noise energy within this critical frequency band, is fundamentally important for assessing whether or not masking is likely to occur.

CBs have been measured both directly and indirectly in a number of marine mammals. In cases where data are available over a wide range of frequencies, critical bandwidth as a proportion of frequency plotted against frequency shows a steep rise at lower frequency and a less pronounced rise at higher frequencies (Figure 3-2). This pattern is also seen in terrestrial mammals. CBs are narrow for odontocetes at high frequencies (>1 kHz) and increase markedly at lower frequencies. This means that at higher frequencies only the noise energy within a narrow band of a signal will be effective in masking it, while at lower frequencies sound energy in a much wider band will cause masking.

Directional Hearing

When noise and a signal arrive at a receiver from different directions, two mechanisms can function to reduce masking. The first relates to the receiving beam pattern of the animal; that is, the extent to which its auditory system is more sensitive to sound on a particular bearing. Normally the direction of greatest sensitivity is ahead, and an attending animal will typically orient toward a sound source so that the absolute level of the sound at the receiver is increased and (provided the noise and signal are on different bearings) the signal-to-noise ratio is also improved. Animals can also determine the direction from which a sound arrives based on cues, such as differences in arrival times, sound levels, and phases at the two ears. The ability that this provides to resolve the signal and noise to different direc-

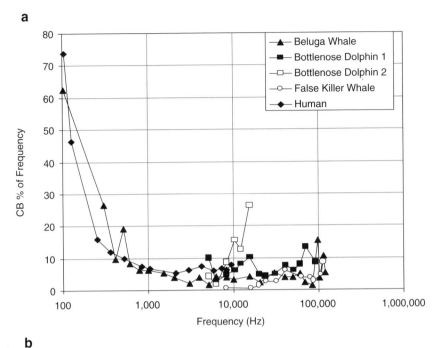

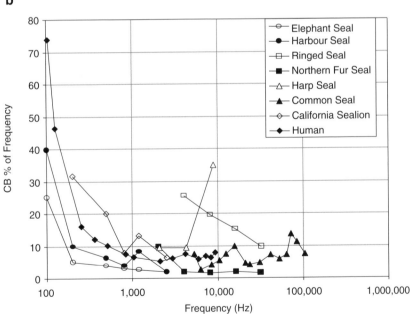

FIGURE 3-2 Critical bands of (a) odontocetes and (b) pinnipeds plotted as a proportion of frequency vs. frequency. SOURCE: Adapted from Wartzok and Ketten (1999).

tions can further reduce masking. Thus, an animal's directional hearing capabilities have a bearing on its vulnerability to masking. Odontocetes have good directional hearing above 1 kHz (Renaud and Popper, 1975), but directional hearing at lower frequencies has been less completely studied. The shielding effects of head structures that are important for both the receiver beam and for causing the sound-level differences at the two ears that contribute to directional hearing are both wavelength dependent. This is reflected by a general trend for a less acute directional hearing ability for lower-frequency sounds. The directivity index (DI) is a measure of the effectiveness of an acoustic receiver in reducing the effects of omnidirectional noise and is expressed as the number of dBs above the signal that omnidirectional noise must rise to mask it. Au and Moore (1984) investigated the DI of a bottlenose dolphin for a signal arriving from ahead and found that it ranged from 10.4 dB at 30 kHz to 20.6 dB at 120 kHz. At these frequencies, then, sounds arriving from ahead, such as echolocation return echoes, will be substantially protected from masking. Directional hearing is less acute in pinnipeds and has not been measured formally in any of the great whales.

Masking of Representative Signals by Realistic Noise

Most studies of masking with captive animals have explored the masking of a very simple signal, typically a pure tone, by broadband noise of constant spectral density (i.e., white noise). In the real world both signals and masking noise are more complex spectrally and temporally, and only a few studies have explored these more realistic scenarios.

The masking effects of noise from oil-spill cleanup vessels on killer whale vocalizations were investigated in a series of experiments conducted with two captive killer whales (Bain and Dahlheim, 1994). Three sets of experiments that varied the characteristics and relative position of the interfering noise were conducted. Boat noise masked all tones below 20 kHz. Masking was reduced when signal and noise sources were separated, and this effect was most pronounced at higher frequencies and greater angles of separation, suggesting the directional hearing ability of the whale was reducing masking. In contrast to the pure tone signal results, when the signal was of biological relevance, that is, killer whale vocalizations, there was little evidence of masking by boat noise.

Concern about interference with beluga whales' communication by icebreaking activity led Erbe and co-workers to explore masking of a beluga call by three different types of icebreaker noise (Erbe, 1997, 2000; Erbe and Farmer, 1998; Erbe et al., 1999). The noise types were ice ramming (primarily propeller cavitation), natural ice cracking, and an icebreaker's bubbler system (high-pressure air blown into the water to push floating ice away from the ship). Bubbler noise was the most effective masker of beluga calls with a critical noise-to-signal ratio (CNSR) of 15.4 dB, followed by ramming noise (CNSR of 18 dB), with natural ice-cracking noise being least effective (CNSR 29 dB). Experiments using trained animals are time consuming and expensive to perform, so a series of software models were designed to exhibit the same masking performance as a beluga whale (Erbe et al., 1999). A neural network model showed the best performance. However, the model was trained and tested using only a single vocalization and three samples of masking noise, and thus may not be robust for other signal and noise combinations. Human performance in masking tests was very similar to that of the beluga whale (Erbe et al., 1999).

Zones of Masking

One way of identifying the potential effects of noise is to determine the areas, or zones of influence, over which particular effects might occur. Richardson et al. (1995) identified four concentric zones with decreasing size and increasing intensity of the signal. The largest zone is that of audibility, followed by responsiveness, then masking, and finally the zone of hearing loss, discomfort, or injury. The outer three zones can be essentially coterminous. If marine mammals attend to barely detectable signals, then any increase in noise may contribute to masking. The zone of masking is defined by the range at which sound levels from the noise source are received above threshold within the CB centered on the signal. A ray-tracing propagation model predicted a zone of masking of beluga whale calls by icebreaker ramming noise of 40 km (Erbe and Farmer, 2000).

Møhl (1981) developed an alternate approach for exploring the significance of different levels of masking noise. He used the sonar equation to show that as the noise increases by a set amount, the range for detecting a signal at a given signal to noise would be reduced by a constant proportion called the range reduction factor (RRF). For example, a 6-dB increase in noise would decrease by half the range for signal detection under transmission loss (TL) determined by spherical spreading, given the same signal-to-noise ratio. Under conditions where TL is given by cylindrical spreading, the range is reduced to one-quarter of its original value. (It is worth noting that in some cases the area over which signals can be detected will be a more appropriate measure than the range, in which case reduced effectiveness resulting from masking will scale in relation to RRF^2.) One attractive feature of Møhl's approach is that it does not require assumptions to be made about the signal-to-noise ratio the animal requires to make detections. It follows directly from the sonar equation that the RRFs resulting from an introduced noise are greater when existing levels of background noise are lower. However, it could be argued that in most cases the appropriate measure of the biological cost of masking relates to the absolute level of signal detection efficiency for the animal in the presence of all noise. In this case, an animal whose auditory efficiency was already reduced by masking from existing higher levels of background noise might be more likely to be adversely affected by an additional masking source than an animal in a quiet environment.

Masking Thresholds

Masking experiments usually measure whether or not any signal can be detected in a particular level of noise. However, detection may not always be the most biologically appropriate measure; in some situations more stringent criteria may apply. Erbe and Farmer (2000) pointed out that relatively low signal-to-noise levels that allow detection might not be sufficient to allow signal recognition. They suggested a higher "recognition threshold" should be considered. An even higher level, an "understanding threshold" may be necessary for an animal to glean all information from complex signals.

Although results from masking experiments are often presented in terms of specific thresholds, it can be more useful to think of masking affecting the probability of correctly detecting a signal. This perspective is particularly appropriate in real-world situations, where levels and spectral characteristics of signal and noise are likely to vary over time.

Strategies to Reduce the Probability of Masking

Marine mammals evolved in an environment containing a wide variety of naturally occurring sounds, and thus they show a variety of strategies to reduce masking. Vocal signals may be designed to be robust to masking effects. Signals can be more easily detected in noise if they are simple, stereotyped, and occur in a distinctive pattern. Signals may also show a high level of redundancy; they may be repeated many times to increase the probability that at least some will be detected. However, these characteristics all minimize the amount of information that a signal can convey. Animals can adapt their behaviors to minimize masking, and it is reasonable to interpret such behavioral changes as an indication that masking has occurred. For example, the vocal output of a beluga whale changed when it was moved to a location with higher levels of continuous background noise (Au et al., 1985). In the noisier environment, the animal increased both the average level and frequency of its vocalizations, as though it were trying to compensate for and avoid the masking effects of, the increased, predominantly low-frequency, background noise levels. Penner et al. (1986) conducted trials in which a beluga whale was required to echolocate on an object placed in front of a source of noise. The animal reduced masking by reflecting its sonar signals off the water surface to ensonify to the object. The strongest echoes from the object returned along a path that was different from that of the noise. This animal's ready application of such complex behavior suggests the existence of many sophisticated strategies to reduce masking effects.

Beluga whales increased call repetition and shifted to higher peak frequencies in response to boat traffic (Lesage et al., 1999). Gray whales increased the amplitude of their vocalizations, changed the timing of vocalizations, and used more frequency-modulated signals in noisy environments (Dahlheim, 1987). Humpback whales exposed to LFA sonar increased the duration of their songs by 29 percent (Miller et al., 2000).

The physiological costs of ameliorating masking effects have not been reported. Although these examples all appear to show animals adapting their vocal behavior to reduce the impact of masking, this does not imply that there were no costs resulting from increased levels of noise. Masking may have been reduced but not eliminated. Costs of the changed behavior, such as increased energetic expenditure on higher-intensity vocalizations and use of vocalizations at suboptimal frequencies cannot be estimated yet.

Critical Research Needs to Understand Effects of Masking

Attempts to assess the masking effects of a particular type of noise in marine mammals are hindered by our poor understanding of how animals make use of the many acoustic cues in the marine environment. Though it is assumed that they attend to, and make use of, each other's communication vocalizations, it is unclear what received levels are necessary to elicit recognition and response to social calls.

The biological implications of signal masking will depend greatly on the function of the signal and the context. In a healthy animal population in which males compete with each other vocally to attract a female, the introduction of masking noise might have little effect because increased noise would disadvantage all males equally. Even if the females' ability to make a mating choice were diminished, they would still be likely to find a mate. In the case of a severely depleted population, the ability of males and females to find each other using acoustic cues could become vital for the well-being of the species. If additional noise reduced acoustic range by masking and effective reproduction were compromised, the consequences for individuals and populations could be very significant (Payne and Webb, 1971; Myrberg, 1980).

How marine mammals make use of the myriad acoustic cues in the marine environment, or the "acoustic scene," is even more poorly understood than masking of communication. Many of these acoustic cues are faint and are thus susceptible to masking by even low levels of noise. While a vocalizing animal may adapt its vocal behavior to compensate for increased levels of masking noise by vocalizing more intensely, changing the emphasized frequency or increasing redundancy, masking of these other acoustic cues cannot be mitigated. A better understanding of the role of passive listening, that is, investigation of the environment through listening without active generation of echolocation pulses, in the lives of marine mammals may well be the most fundamental research need for assessing masking impacts. Detailed field research involving fine-scale behavioral observations linked to sensitive real-time acoustic monitoring will be required to gain any appreciation of how marine mammals utilize these low-level noises.

To investigate the occurrence of masking in the real world, field projects could be designed to study behavioral changes, thought to be indicative of masking (such as the strategies to avoid masking outlined earlier), and behavioral performance in situations with different levels of background noise could be monitored (see also recommendations in NRC, 2000; Appendix D). Measures of feeding rates and hunting success, mate-searching behavior, and predator avoidance would be necessary to elucidate whether masking effects were likely to affect the survival or reproduction of the individual and ultimately impact populations.

HABITUATION, SENSITIZATION, AND TOLERANCE OF MARINE MAMMALS TO MARINE NOISE

Habituation to repeated presentations of a signal that is not associated with physical discomfort or overt social stress is a common adaptive feature of sensory systems that predates the evolution of mammals. It is not surprising that marine mammals show habituation to many signals that initially cause an overt reaction. To demonstrate habituation, the same signal needs to be presented to the same individual repeatedly and the response of that individual charted over the sequential presentations. Such a demonstration in marine mammals is rare. Instead, habituation is inferred by the changes in the response of animals of the same species in the same area over time. This assumes that although the individuals are unidentified in the group, there is consistency in group composition over the course of the study. A second-order inference of habituation can also be made by comparing the reactions of individuals of the same species from two different areas to the same stimulus, the stimulus being one to which animals in one area have been exposed previously, whereas animals in the other area are assumed naive with respect to this particular stimulus.

Some of the clearest evidence of habituation comes from attempts to use sound sources to keep marine mammals away from an area or a resource (Jefferson and Curry, 1994). Acoustical harassment devices (AHDs) have been used in an attempt to keep pinnipeds away from aquaculture facilities or fishing equipment. AHDs emit tone pulses or pulsed frequency sweeps in the 5-30 kHz range at source levels up to 200 dB re 1 μPa at 1 m. Although initially effective, over time some of the devices became less able to deter harbor seals (*Phoca vitulina*), presumably because of habituation (Mate and Harvey, 1987) but also because of a change in seal behavior in which the animals spend more time swimming with their heads out of the water when they are in intense sound fields. Seals and California sea lions (*Zalophus californianus*) even habituate to "seal bombs" that can have peak sound pressure levels of 220 dB re 1 μPa at 1 m (Mate and Harvey, 1987; Myrick et al., 1990). Harbor porpoises (*Phocoena phocoena*) habituate to pingers placed on gillnets in an attempt to reduce the porpoise bycatch. The probability of porpoises being within 125 m of a pinger decreased when the pinger was first activated, but within 10-11 days had increased to equal the control (Cox et al., 2001).

Watkins (1986) summarized 25 years of observations of whale responses near Cape Cod to whale-watching boats and other vessels. Minke whales (*Balaenoptera acutorostrata*) changed from frequent positive interest to generally uninterested reactions. Fin whales (*B. physalus*) changed from mostly negative to uninterested reactions. Humpbacks (*Megaptera novaeangliae*) changed dramatically from mixed responses that were often negative to often strongly positive reactions, and right whales continued the same variety of responses with little change. Gray whales wintering in San Ignacio Lagoon are less likely to flee from whale-watching boats later in the season than they are shortly after arriving in the lagoon (Jones and Swartz, 1984). In all these examples, factors in addition to habituation could have contributed to the observed changes.

In contrast to habituation, which results from repeated presentations of an apparently innocuous stimulus, sensitization is the result of prior presentation of a stimulus that either by itself or in conjunction with another action results in a negative experience for the animal. In sensitization, responses at subsequent presentations are more marked than are the responses at the initial presentation. Northern fur seals (*Callorhinus ursinus*) showed little initial reaction to a ship, but if that ship were subsequently used in seal hunting, the seals avoided it at distances up to a mile (H. Kajimura, in Johnson et al., 1989). Walruses hauled out on land are more tolerant of outboard motorboats in years when they are not hunted from such craft than they are in years when these boats are used in walrus hunts (Malme et al., 1989). Bottlenose dolphins that had previously been captured and released from a 7.3-m boat would flee when that boat was more than 400 m away, whereas bottlenose dolphins that had not been captured by the boat often swam quite close to it (Irvine et al., 1981). All the reported cases of sensitization are the result of conditioning: the pairing of a given stimulus with a significantly negative experience.

Animals will tolerate a stimulus they might otherwise avoid if the benefits in terms of feeding, mating, migrating to traditional habitat, or other factors outweigh the negative aspects of the stimulus. Already noted is the case of bowhead whales on spring migration, where they needed to use the one available lead in the ice cover to continue on their eastward migration and passed through a sound field with projected drilling ship sounds at levels of 131 dB re 1 µPa (Richardson et al., 1991). Bowheads also return to the same areas of the Canadian Beaufort Sea year after year even though seismic surveys occurring at the same time are an annual feature of these areas (Richardson et al., 1987). Whether there are particularly dense concentrations of prey in these areas or whether the bowheads' response is simply historical philopatry is unknown.

In at least one case, a source that did not elicit a fleeing response turned out to be capable of causing damage. Humpback whales in Newfoundland remained in a feeding area near where seafloor blasting was occurring. The humpbacks showed no behavioral reaction in terms of general behavior, movements, or residency time. In fact, residency time was greater in the bay closest to the blast site than it was in other bays of equivalent size and productivity nearby. Estimated peak received levels during blasting were approximately 153 dB re 1 µPa with most of the sound energy below 1,000 Hz (Todd et al., 1996). Two humpback whales found dead in fishing nets in the area had experienced significant blast trauma to the temporal bones (Ketten et al., 1993).

ACOUSTICALLY INDUCED STRESS

Acute responses to sounds may be difficult to quantify, but they are much more tractable to investigation than are responses to repeated or chronic sounds. Sounds resulting in one-time acute responses are less likely to have population-level effects than are sounds to which animals are exposed repeatedly over extended periods of time. Long-term population effects will have the greatest impact on marine mammal species.

Long-term effects of ocean sounds can include the transformation of TTS to permanent threshold shift and an increase in occurrence of pathological stress. Stress can be defined as a perturbation to homeostasis. So long as the perturbation is within the range the physiological system is capable of handling, is of short duration, and is not continually encountered, homeostasis is restored through an adaptive stress response. However, when the perturbation is frequent, outside the normal physiological response range, or persistent, the stress response can be pathological.

Stress can induce secretion of corticotrophin releasing factor (CRF) from the hypothalamus. CRF promotes the release of glucocorticoids and catecholamines, which modulate the immune response and can lead to changes in the response to infectious, neoplastic, allergic, inflammatory, and autoimmune diseases (Webster et al., 1977). Chronic stress can also suppress reproduction (Rabin et al., 1988), inhibit growth (Diegez et al., 1988), and alter metabolism (Mizrock, 1995).

Although stress-induced pathologies have been hard to identify in free-ranging marine mammals, based on work with terrestrial mammals, it is likely that marine mammals would experience the same responses. The stress caused by pursuit and capture activates similar physiological responses in terrestrial mammals (Harlow et al., 1992) and cetaceans (St. Aubin and Geraci, 1992). One of the first recognized effects of chronic stress was the hypertrophy and hyperplasia of the adrenal cortex and medulla (Selye, 1973). Some possibly stress-induced adrenal pathologies have been observed in marine mammals. Harbor porpoises that died of chronic causes were more likely to exhibit adrenocortical hyperplasia than were ones that died of acute causes (Kuiken et al., 1993). Mass-stranded Atlantic white-sided dolphins had adrenal cysts, which were possibly stress related (Geraci et al., 1978). Both adrenocortical hyperplasia and cysts were observed in stranded beluga whales with the incidence and severity of the lesions increasing with age, although the authors could not attribute the adrenocortical changes to chronic stress, in contrast to normal aging (Lair et al., 1997).

Controlled laboratory investigations of the response of cetaceans to noise have shown cardiac responses (Miksis et al., 2001) but have not shown any evidence of physiological effects in any of the blood chemistry parameters measured. Beluga whales exposed for 30 min to 134-153 dB re 1 μPa playbacks of noise with a synthesized spectrum matching that of a semisubmersible oil platform (Thomas et al., 1990) showed no short-term behavioral responses and no changes in standard blood chemistry parameters or in catecholamines. Preliminary results from exposure of a beluga whale and bottlenose dolphin to a seismic watergun with peak pressure of 226 dB re 1 μPa showed no changes in catecholamines, neuroendocrine hormones, serum chemistries, lymphoid cell subsets, or immune function (Romano et al., 2001).

Among terrestrial mammals, a bank of blood indicators is a more reliable measure of stress across species or within species and across time (Hattingh and Petty, 1992). In cetaceans, Southern et al. (2001) and Southern (2000) are attempting to develop microassays to detect in skin samples from free-ranging cetaceans changes in a suite of 40 stress-activated proteins.

Although techniques are being developed to identify indicators of stress in natural populations, determining the contribution of noise exposure to those stress indicators will be very difficult but important to pursue in the future when the techniques are fully refined.

NEW RESEARCH TOOLS TO UNDERSTAND MARINE MAMMAL BEHAVIOR

Any real understanding of long-term and cumulative effects of noise on marine mammals will require the development and refinement of a number of new research instruments. Ideally, sound pressure level should be recorded as the animal receives it and the vocalizations of the animal also need to be recorded in real time along with as many movement parameters and physiological parameters as possible. Recently several new tags have been developed that incorporate some of these features. Researchers working on northern elephant seals, *Mirounga angustirostris*, have developed acoustic recording packages that include a hydrophone and temperature and depth sensors (Burgess et al., 1998) or a digital audio recorder with a time-depth recorder and a time-depth-velocity recorder (Fletcher et al., 1996) in a package that can be placed on juvenile seals. The tags record received sound, seal swim strokes, and during quiet intervals at the surface both respiration and heartbeats. Cetacean researchers further developed these concepts into digital sound recording tags that record onto solid-state memory received signal levels, animal vocalizations, pitch roll and orientation, and depth (Burgess, 2001; Johnson et al., 2001; Madsen et al., 2002). Three-dimensional tracks of the whale's movements can be reconstructed from the recorded data. These tags are typically applied with suction cups so although they provide a lot of data, it is only for a short time period. Another tag places a suction-cup hydrophone on a dolphin to record heartbeats. This has been tested so far on captive animals where the dolphin showed significant heart rate accelerations in response to playbacks of conspecific vocalizations compared to baseline rates or to playbacks of tank noise (Miksis et al., 2001). Finally, radio tags need to be developed that remain attached for several years and transmit only on a programmed cycle or in response to a query signal. For most marine mammal species, the difficulty in identifying individual animals rapidly and reliably makes it very difficult to follow animals for long periods of time to determine cumulative effects. Borggaard et al. (1999) were able to follow individually identified minke whales over four years and noted that this provided a more sensitive means of assessing impacts of industrial activity than did abundance and distribution measures. At a minimum, animals must be identified and observed preexposure, during exposure, and postexposure for a sufficient number of repetitions and for a sufficient period of time to be able to make any reasonable statements on the effect of the exposure on a given animal and potentially on the population. Without these data, we will simply continue to collect disparate observations of transient behavior, which tell us little about the impact of anthropogenic noise on marine mammals.

MARINE ECOSYSTEM IMPACTS OF NOISE

While the focus of the concern regarding the impact of marine ambient sounds is on mammals, mammals make up only a tiny fraction of all marine species. Moreover, other marine organisms, fishes and invertebrates, are critical components of the food chain for marine mammals (and terrestrial mammals, including humans), and any impact on these organisms, or their eggs and larvae, could have significant impact on mammals.

The data on the impact of sound on fishes are very limited and nonexistent for reptiles and invertebrates. A few studies that suggest that exposure to high-level pure tones for an hour or more will damage the sensory cells of the ears of a few species (one freshwater and one marine; Enger, 1981; Hastings et al., 1996), although the extent of damage is limited and only occurs after several hours of continuous exposure. Moreover, there is evidence that fish will recover from drug (aminoglycoside antibiotic) induced hair cell damage over a period of several weeks (Lombarte and Popper, 1994).[2] At the same time, during a recovery period of several weeks, fish are without a full set of sensory cells and so they may not be able to detect predators and prey, and thus have a substantially decreased chance for survival.

There are significant caveats on the fish noise-exposure studies [see Hastings et al. (1996) for a full discussion]. First, the studies were done with just a few species, and only Enger (1981) used a marine species, so it is not clear if these data can be extrapolated to other species. Second, the exposure in all of the studies was for long periods of time and to pure tones. Since most anthropogenic noise is likely to be of short duration, extrapolation from long-term continuous exposure to short-term or pulsed exposure may be inappropriate. Third, the animals in these experiments were confined near the sound source. Since fish are free to move around, it might be expected that they would move away from an intense sound.

Another issue is the sound levels used in the few fish studies. In both studies, sounds were 90-140 dB above threshold (about 180 dB re 1 μPa).

Perhaps a more significant study is one on the impact of air-guns on the ears of a variety of Australian marine fishes. In this study, fish were exposed to the sound of a small air-gun and the ears collected for analysis of inner ear hair cell damage (McCauley et al., 2000, 2003). The results show that exposure to air-guns with a maximum received level of 180 dB re 1 μPa over 20-100 Hz causes major damage to sensory cells of the ear of at least one species. Despite a number of caveats to these results, they suggest air-guns damage sensory hair cells in fishes. While similar studies have not been done with marine mammals, one must question whether these results could also have implications for marine mammals exposed to air-guns, particularly since the hair cells in fishes and marine mammals are so similar to one another.

There are also data that suggest that there may be significant impacts on fish behavior from air-guns, and perhaps from other sound sources. Several studies suggest that intense sounds may result in fish moving from

---

[2]While the sensory cells of the ears of fishes and marine mammals are the same, regeneration of damaged cells does not occur in mammals.

an area for extended periods of time. For example, Engås et al. (1996) showed a significant catch decrease in a fishing area after use of air-guns, suggesting that fish moved from the ensonified area and only returned days later. There is also some evidence low-frequency noise produced by fishing vessels and their associated gear may cause fish to avoid the vessels (Maniwa, 1971; Konagaya et al., 1980). While all of these data need replication, they do suggest that sounds may change the behavior of fish. Movement of fish from a feeding area of marine mammals (or fishing areas for humans) could have an adverse impact on the higher members of a food chain and therefore have long-term implications despite the fish themselves not being killed or maimed.

Another concern is the impact of high-level anthropogenic sounds on overall behavior. Since many species of fish use sound for attracting mates and for other behaviors, any masking of these sounds could alter behavior. Increased environmental sounds in the vicinity of coral reefs may have a substantial impact on settling of larval fish on the reefs. Larval reef fish of many species spend part of their lives offshore and away from reefs, and then need to find a reef where they will live for the remainders of their lives (Leis et al., 1996). Recent evidence suggests that at least some larval fish are likely to use the reef sounds to find the reefs and that the fish will go to regions of higher-level sounds (Tolimieri et al., 2003). Thus, if there are intense offshore sounds, larval fish may be confused and not be able to find the reef. Alternatively, such sound may mask reef sounds, again preventing larval fish from finding the reef.

Potentially, anthropogenic sounds can have effects on marine life at a number of different levels, from short-term effects on individuals to long-term effects on populations and even species. Effects that can be dramatic, even lethal, at the level of the individual may have negligible consequences at the population level if, for example, small numbers of a large healthy population are affected. Conversely, effects that may seem insignificant for the well-being of individuals could have important conservation consequences for populations that are depleted and under stress. For example, a decrease in feeding rate that might equate to a year's delay in attaining sexual maturity, a small increase in infant mortality, or a slightly shorter life span may not be overly significant to an individual animal but could mark the difference between extinction and recovery for a critically endangered species. It is important to emphasize that whether or not a particular impact could be of conservation significance will depend on the status of the population; thus, the conservation significance of particular impacts must be assessed on a case-by-case basis. While much legislation and scientific work focuses on conservation goals, it is important to recognize that the well-being and welfare of individual wild animals is also a concern for many members of the public and harassment of any individual marine mammal is prohibited by the Marine Mammal Protection Act.

4

Modeling and Databases of Noise in the Marine Environment

INTRODUCTION

The task statement for this committee states: "The study will review and identify gaps in existing marine noise databases and recommend research needed to develop a model of ocean noise that incorporates temporal, spatial, and frequency-dependent variables." This chapter describes current acoustic models and extant databases of underwater noise and discusses efforts to model noise effects in marine mammals. High-quality, well-documented databases are essential for model validation and further model development and should contain information on the various environmental and biological factors that control the impact of noise on marine mammals. Gaps that must be filled to model the impact on marine mammals are identified for both models and databases. However, as with all models of the physical world, uncertainties in parameters and approximations in the modeling techniques are inevitable and must be accounted for using statistically valid means when interpreting the model predictions.

ACOUSTIC MODELING OF THE MARINE ENVIRONMENT

Noise in the ocean is usually broken into two broad categories based on the type of source. The first type of noise is generated by a single, identifiable, and usually close source of noise, such as an air-gun array or one or more marine mammals or other biological sources. The second type is generated by multiple indistinguishable sources of noise, such as vessels in a shipping lane and whitecaps. Some important parameters for characterizing the effects from single sources are frequency, source level, pattern of amplitude versus time (time series), directionality of radiation or beam pattern, and distance from the source. Effects from multiple unidentified sources are primarily characterized by frequency, directionality, and level at the receiver. To underwater acousticians, the term "ambient noise" refers to the second type of noise from multiple and unidentifiable sources as stated in Chapter 1.

Models are used to assess the interactions of sound fields created by multiple sources, propagation through space and time, and interactions with marine mammals. The term "models" refers to a variety of tools, including empirical fits to measured data, such as the Wenz curves, computer simulation models, and numerical models, which can be either physics or empirical based. Physics models rely on known relations such as those expressed in Equations 1-1 to 1-5. Empirical models are based on observed data rather than underlying physics. In many cases the dominant mechanisms of natural sources of ocean ambient noise, for example, those associated with wind-generated noise, have not yet been conclusively identified. Therefore, physics-based approaches that incorporate actual source mechanisms are still in their infancy in underwater acoustics. In contrast, empirical models such as the Knudsen curves (Knudsen et al., 1948) and the Wenz curves (Wenz, 1962) have been extremely successful; they remain the basis of standardized noise spectra used by the U.S. and British navies.

The first part of this chapter describes current acoustic models and efforts to model underwater noise effects on marine mammals. Gaps that must be filled to model the effects of noise on marine mammals are identified in modeling efforts and current databases.

Modeling Single Sources of Noise

Some ocean noise can be traced to a single identifiable source. High-quality models exist to predict the time series of the received signal from a source of specified directivity and given transmitted signal time series. Propagation models utilize bathymetric databases, geoacoustic information, oceanographic parameters, and boundary roughness models to produce estimates of the acoustic field at any point far from the source (see Glossary for definitions). The quality of the estimate is directly related to the quality of the environmental information used in the model. For example, in continental shelf waters, geoacoustic parameters such as compressional sound speed, attenuation, and sediment density can significantly affect the acoustic propagation. Variability introduced in these parameters can substantially affect model predictions; propagation loss can be incorrect by as much as 20 dB as a result of inaccurate geoacoustic parameters.

There are four main categories of acoustic propagation models prima-

TABLE 4-1 Propagation Models and Other Information Available from the Current Contents of the Ocean Acoustics Library at SAIC

Category	Models
Parabolic equation	FOR3D, MMPE, PDPE, RAM, UMPE
Normal modes	AW, COUPLE, KRAKEN, MOATL, NLAYER, WKBZ
Wavenumber integration	OASES, RPRESS, SCOOTER, SPARC
Rays	BELLHOP, HARPO, RAY, TRIMAIN
Other	Related modeling software and data sets to support oceanographic and acoustical analyses

SOURCE: *http://oalib.saic.com*; Etter (2001). Reproduced courtesy of Academic Press/ Elsevier Ltd.

rily used in underwater acoustics: parabolic equation (PE), normal mode, wavenumber integration, and ray models. Each of these different categories represents a different approach to simplifying either the acoustic wave equation (the fundamental mathematical equation that contains all the basic physics of sound propagation) or the model of the environment, or both. Simplification is required in order to allow computer codes to be constructed and to make them computationally efficient. Accuracy of all four model types is dependent on the frequency of sound being modeled and the environmental characteristics. In general, the PE is used for range-dependent environments at frequencies below 1,000 Hz. Normal mode models can be significantly more efficient for modeling in some environments at frequencies below 1,000 Hz. The accuracy of most normal mode models is limited in strongly range-dependent environments such as the continental shelf and slope. Wavenumber integration is usually limited to frequencies below 1,000 Hz and typically is limited to range-independent environments, although this approach recently has been extended to range-dependent environments. Ray codes are accurate and efficient for most environments but are limited to frequencies usually above 1,000 Hz. For all the models mentioned, azimuthal coupling resulting from three-dimensional medium variability (i.e., the transfer of acoustic energy propagating in one azimuthal direction into energy propagating in a different azimuthal direction) is not modeled and is considered less important than the effects of environmental uncertainty. Many propagation models are available to the public (Table 4-1). Examples of transmission loss computed using the MMPE model show the complexities of the propagation process, as well as the substantially reduced sound level at 3,000 Hz, when compared to those of 200 Hz, for longer ranges (Figure 4-1). The latter behavior is due to the effect of increased absorption at higher frequencies (cf. Figure 1-2 and Table 4-2).

112

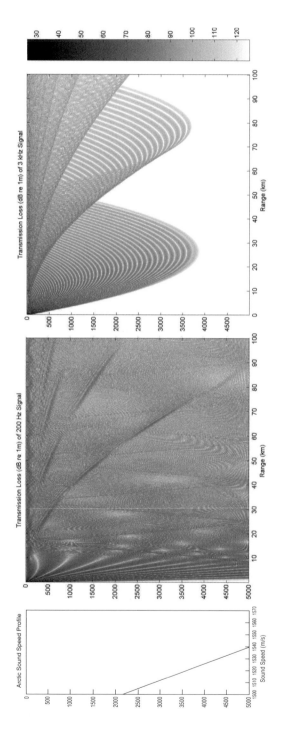

FIGURE 4-1 Transmission loss in different oceanic regimes as predicted by the MMPE model at both 200 Hz and 3 kHz. (a) Arctic, with a point source depth of 50 m;

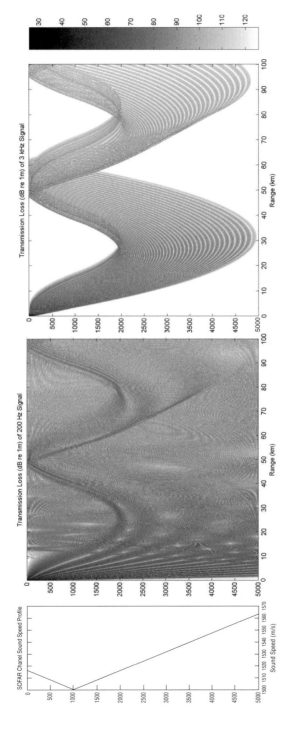

FIGURE 4-1 (b) the SOFAR channel at mid-latitude and a point source depth of 1,000 m;

114

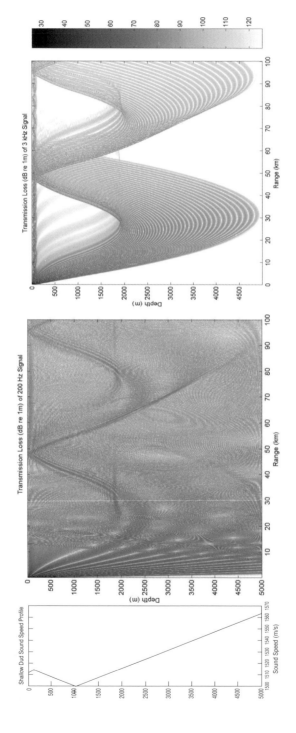

FIGURE 4-1 (c) a surface duct in the mid-latitude ocean.

TABLE 4-2 Absorption by Seawater for Two Frequencies for a Range of 1,000 km[a]

Noise Source	Typical Frequency	Absorption Loss at 1,000 km	Absorption Loss at 10 km
Shipping	100 Hz	2 dB	0.002 dB
Wind	1,000 Hz	60 dB	0.6 dB

[a]Note that these losses are in addition to geometrical spreading and scattering losses.

Modeling Distributed Sources of Noise

The Wenz curves are used to predict or model the noise level from unidentifiable sources (Plate 1; Wenz, 1962). These curves provide the noise spectrum level that a theoretical ideal receiver receives, given in decibels referenced to 1 $\mu Pa^2/Hz$. An ideal receiver has an omnidirectional reception sensitivity—in other words, its sensitivity does not vary with direction. Ambient noise is a random quantity, meaning that a given realization of the noise time series is unpredictable. However, statistical characteristics of the time series such as its variance are predictable (see Glossary). Low-frequency noise is usually much higher level than high-frequency noise due because of the character of the noise sources themselves and also as a result of the frequency dependence of sound absorption in the ocean, as described below. Typically, the property of the noise that is modeled is its pressure spectral density level. A spectrum and spectral density are frequency catalogues of a time-varying signal. The pressure spectral density of ambient noise, modeled as a random process, is the variance per hertz of the pressure time series ($\mu Pa^2/Hz$). For a deterministic process, the pressure spectral density is the mean squared pressure per hertz (see Glossary). Below 10 Hz, microseisms caused by the nonlinear interaction of ocean surface waves are the dominant source of ocean noise. Earthquakes also contribute intermittently.

Between 10 and 200 Hz distant shipping is the largest contributor to the noise spectrum level (Wenz, 1962). From 200 Hz to 80 kHz, wind-generated breaking waves are the primary contributor to ambient noise. These levels are dependent on wind speed, and data validate the model (Felizardo and Melville, 1995). These ambient noise spectra use 1-Hz bands, while studies of noise masking in mammalian ears has typically found that one-third-octave bands are good models for these ears. For example, the one-third-octave band centered at 50 Hz runs from approximately 45 to 56 Hz. To convert a 1-Hz band level to a one-third-octave band, 10 times the logarithm of the bandwidth is added to the 1-Hz band level. For the one-third-octave band centered at 50 Hz, this translates to about a 10-dB increase. At the one-third-octave band centered at 3,000 Hz, the difference between it and the 1-Hz band is approximately 28 dB.

Ambient noise from distant sources is affected by the environment. Noise absorption by seawater is strongly dependent on frequency, effectively limiting the distance high-frequency sounds propagate (Figure 1-2). Absorption causes a decrease in received signal levels (i.e., an increase in transmission loss), which occurs in addition to the decrease produced by geometrical spreading effects, as discussed in Chapter 1.

The absorption of shipping noise in the 1-Hz band, centered at 100 Hz, is approximately 0.002 dB per km. In other words, 1,000 km from a source of 100 Hz, the attenuation loss is about 2 dB in addition to the geometrical spreading losses. For higher-frequency sound, such as that generated by wind at 1,000 Hz, the absorption factor increases to approximately 0.06 dB per km. At a distance of 1,000 km from a 1,000-Hz source, the attenuation loss is about 60 dB (Table 4-2). For distant sources of ambient noise, frequency largely determines the region over which these sources can be important. Ships contribute to ambient noise at ranges of hundreds of kilometers, while wind noise contributes to ambient noise for distances of kilometers.

It should be noted that the majority, if not all, the models for oceanic ambient noise have been developed for and supported by Navy sponsors. Appendix C provides a summary of underwater acoustic noise models. This summary is not meant to be all-inclusive but rather to indicate some of the better-known and more heavily used examples. Over the decades since World War II, naval sonar systems, starting from simple transducer units, have increased in complexity. Initially, the sonars operated in an omnidirectional mode and required only a knowledge of ambient noise as seen with that sensor. As the systems acquired more and more directional discrimination to help localize targets, knowledge of ambient noise directionality was also required. The initial attempts to define and measure noise directionality were confined to studies of the variations either in the vertical direction only or the azimuthal direction only. Later, as the sonar arrays became even more spatially discriminating, beam noise estimates were required where both horizontal and vertical limits were used.

From the perspective of an omnidirectional system, the Wenz curves would be the model required. The summary in Appendix C, then, goes on from that point, with the ambient noise models where there are listed a number of directional models with respect to either the horizontal or vertical plane(s). The beam noise statistics category provides those models that describe beam noise properties. For more details the reader is directed to excellent texts devoted to underwater ambient noise, modeling, and mechanisms (Etter, 1996; Applied Acoustics, 1997).

Models such as ANDES, CNOISE, and RANDI provide predictions of the geographic, seasonal, frequency, and directional dependence of ambient

Color Plates

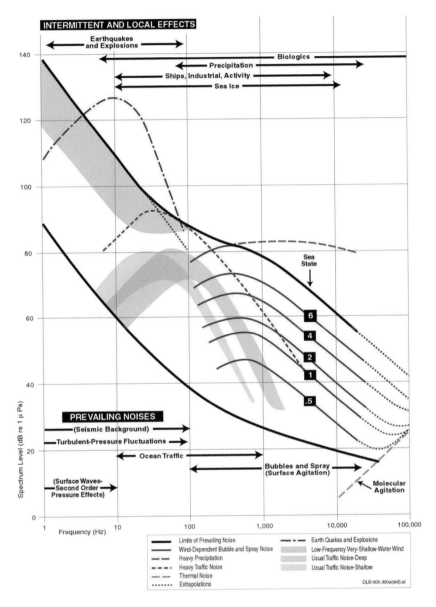

PLATE 1 Wenz curves describing pressure spectral density levels of marine ambient noise from weather, wind, geologic activity, and commercial shipping. (Adapted from Wenz, 1962.)

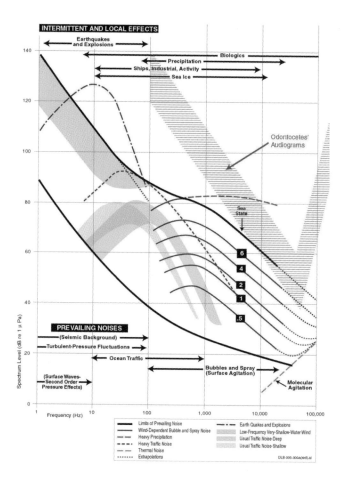

PLATE 2 Range of odontocete audiograms superimposed on the background noise levels. The Wenz curves describe relative levels of marine ambient noise from weather, wind, geologic activity, and commercial shipping. The audiograms are pressure spectral levels with units of dB re 1 μPa, whereas the Wenz noise curves are those of pressure spectral density having units of dB re 1 μPa^2/Hz. (Actually, Wenz collected ambient noise spectra for various frequency bandwidths and converted their levels to a 1-Hz ["1-cps"] bandwidth [Wenz, 1962].) The Wenz curves can be converted into spectral levels for a frequency band of interest by integrating the spectral density levels across that frequency band of interest, after first converting from logarithmic to linear units of μPa^2/Hz. The comparison of spectral and spectral density levels shown assumes the bandwidth of integration is 1 Hz; the spectral density level is equivalent to the spectral level for a 1-Hz-wide bandwidth. From a biological perspective, the appropriate frequency band over which to integrate noise spectral densities is determined by the frequency discrimination capabilities of the animals' hearing. This figure illustrates the similarity in the frequency dependence of naturally occurring wind noise and marine mammal hearing capability and indicates how the frequency content of other noise sources (e.g., shipping) relate to this hearing capability. (Unpublished abstract JASA, 2001; adapted from Wenz [1962] and presented at Acoustical Society of America, December 2001.)

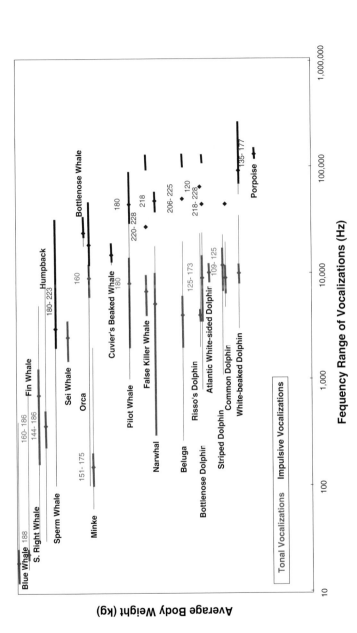

Fequency Range of Vocalizations (Hz)

PLATE 3 Representative vocalizations of marine mammals by average adult body weight. Tonal vocalizations are plotted in red; impulsive vocalizations are shown in blue. The thicker lines indicate the frequencies near maximum energy, and the thin lines indicate the total range of frequencies in vocalizations. Numbers above the line indicate measured source levels in dB re μPa at 1m. Body weight data are taken from Table 10.1 in Boness et al. (2002); vocalization data are summarized from Table 4.1 in Warzok and Ketten (1999) updated with additional information from Hooker and Whitehead (2002), Frantzis et al. (2002), Møhl et al. (2000), and Rasmussen et al. (2002).

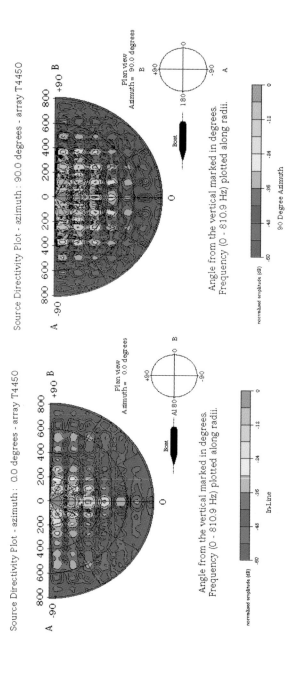

Source Directivity Plot - azimuth : 0.0 degrees - array T4450

Source Directivity Plot - azimuth : 90.0 degrees - array T4450

Angle from the vertical marked in degrees.
Frequency (0 - 810.9 Hz) plotted along radii.

Angle from the vertical marked in degrees.
Frequency (0 - 810.9 Hz) plotted along radii.

PLATE 4 Air-gun array directivity. The amplitude and frequency content of air-gun signals are dependent on the downward angle and the horizontal azimuth of the emitted sound. The frequency response of the 4550 air-gun array signal is shown as a function of direction. The radial lines on each plot represent emission angles from the vertical (0, 30, 60, 90), and the concentric half-circles represent increasing frequencies. Cold colors indicate lower amplitudes, and the hot colors indicate higher amplitudes. The plot on the left shows the response along the inline axis of the array (fore to aft), and the plot on the right shows the cross-line response (port to starboard). In both instances it can be observed that, in general, there are less high frequencies in the emissions away from the vertical. Courtesy of Philip Fontana, Veritas DGC.

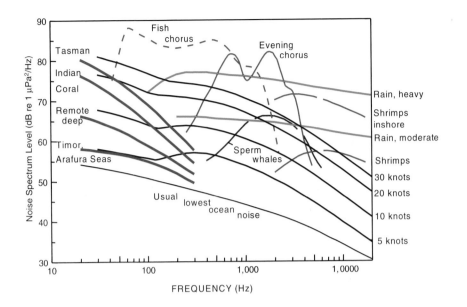

PLATE 5 Ambient noise prediction curves for Australian waters.
SOURCE: Cato, 2001.

50 Hz Winter Predicted Wind Values

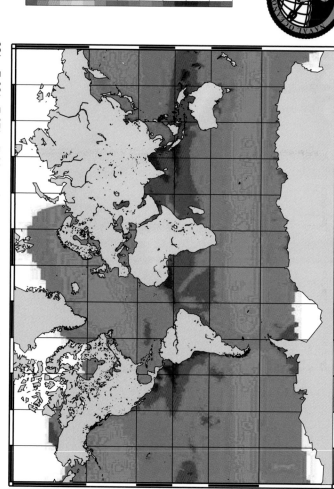

dB//μPa²/Hz

PLATE 6a-d An example wind noise estimate generated by the U.S. Naval Oceanographic Office. Wind speeds for the center months of the season were extracted from the OAML Surface Marine Gridded Climatology (SMGC) database at a 1-degree grid spacing. Wind speeds were converted to noise levels using Wenz curves. Predictions of noise in the absence of other noise producers are shown for the 50-Hz frequency band. (NOTE: Summer includes July-September; winter includes January-March.)

3,500 Hz Winter Predicted Wind Values

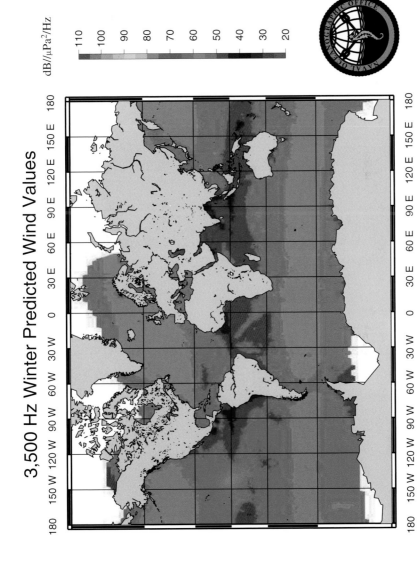

dB/μPa^2/Hz

110
100
90
80
70
60
50
40
30
20

PLATE 6b

50 Hz Summer Predicted Wind Values

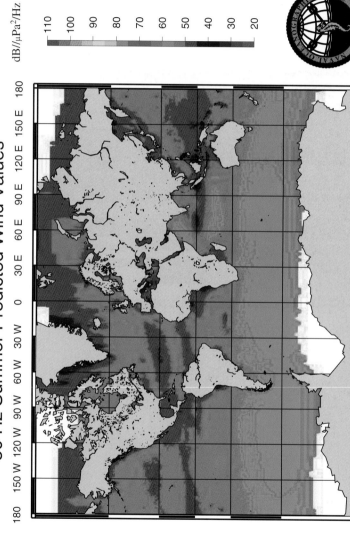

dB//μPa²/Hz

- 110
- 100
- 90
- 80
- 70
- 60
- 50
- 40
- 30
- 20

PLATE 6c

3,500 Hz Summer Predicted Wind Values

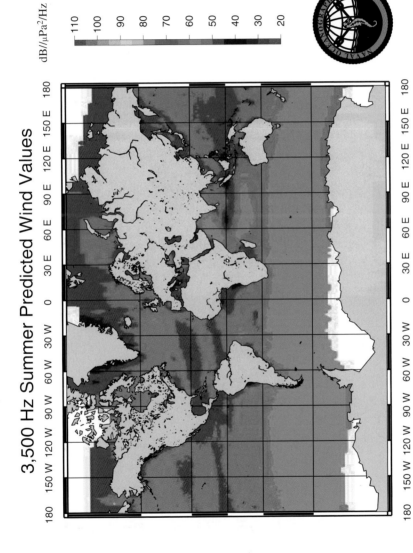

dB//μPa²/Hz

110
100
90
80
70
60
50
40
30
20

PLATE 6d

Mean of 50 Hz Winter Measured Values

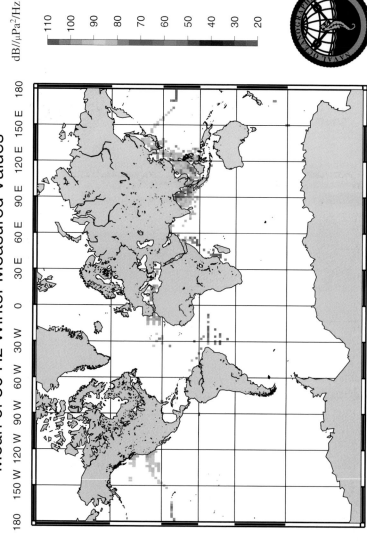

dB/1 μPa²/Hz

110
100
90
80
70
60
50
40
30
20

PLATE 7a-d Mean ambient noise measurements for the world's oceans. The mean values of measured data are displayed in units of dB/1 μPa²/Hz, data are given for 2-degree cells where more than one data point exists. Unshaded areas have fewer than two datapoints in the NAVOCEANO database. The mean of the measured decibel values was calculated and reported at the center of the cell. Data are displayed for winter (January–March) (a) 50 Hz and (b) 3,500 Hz and summer (July–August) (c) 50 Hz, and (d) 3500 Hz.

Mean of 3,500 Hz Winter Measured Values

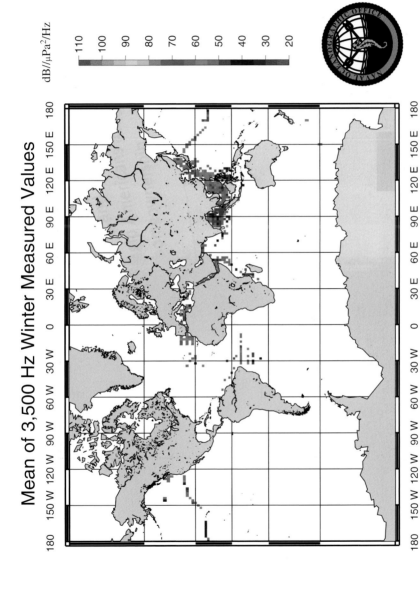

dB//µPa²/Hz

110
100
90
80
70
60
50
40
30
20

PLATE 7b

Mean of 50 Hz Summer Measured Values

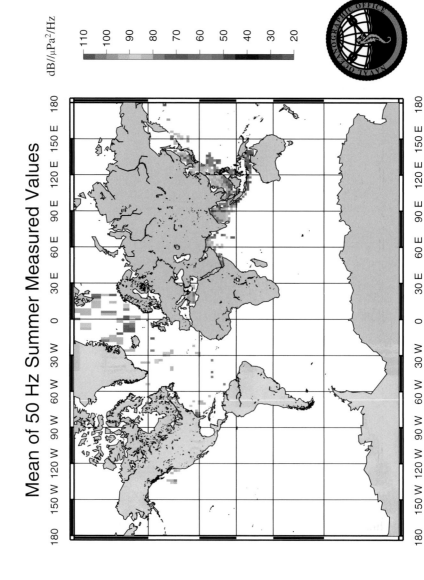

dB//μPa²/Hz

110
100
90
80
70
60
50
40
30
20

PLATE 7c

Mean of 3,500 Hz Summer Measured Values

dB//µPa²/Hz

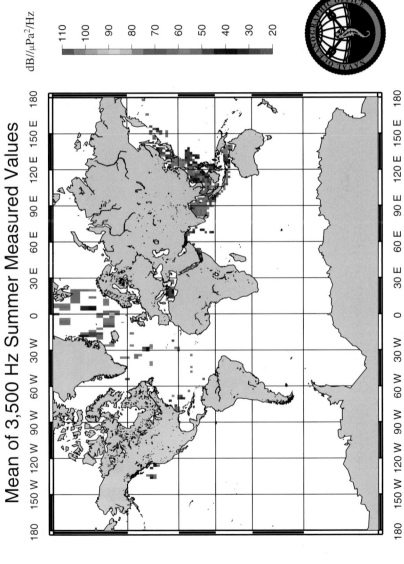

PLATE 7d

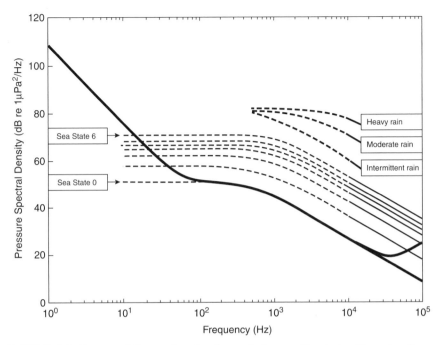

FIGURE 4-2 Output of Comprehensive Acoustic System Simulation/Gaussian Ray Bundles (CASS/GRAB) Model. The thick solid curve shows the base level with no shipping noise, a sea state of 0, and no rain. The seven parallel dashed curves from 10 Hz to 100 kHz show the surface agitation component only for sea states 0 through 6 (in ascending level). The three dashed curves from 550 Hz to 15.5 kHz represent the rain component for intermittent (lower curve), moderate (middle curve), and heavy (upper curve) rain. SOURCE: Naval Undersea Warfare Center Division.

noise from multiple unidentified sources such as distant shipping and wind. These models include shipping density statistics, wind-speed databases based on meteorological models, state-of-the-art propagation models, and oceanographic databases. The models are usually used for sonar performance prediction and maintained by the world's navies. No existing model is capable of predicting the effects of distributed noise sources on marine mammals.

An omnidirectional ambient noise model is included as part of the sonar simulation model CASS/GRAB (Comprehensive Acoustic System Simulation/Gaussian Ray Bundles; Weinberg and Keenan, 1996; Weinberg et al., 2001). The CASS/GRAB model, approved by the Oceanographic and Atmospheric Master Library (P.C. Etter, 1996, 2001) was developed at the Naval Undersea Warfare Center (formerly NUSC) using empirical fits to ambient noise measurements (Figure 4-2). The model accounts for "ocean turbulence," dominant in the 1-10-Hz band, shipping noise prevalent from 10 to 500 Hz, "surface agitation" from 500 Hz to 100 kHz, and thermal noise at frequencies greater than 100 kHz. Noise from rain also is included in the 550-Hz to 15.5-kHz band.

Dynamic Ambient Noise Prediction System

The Dynamic Ambient Noise Prediction System (DAPS) is the most recent development in the succession of U.S. Navy ocean ambient noise models. It is composed of three modules:

- Historical Vessel Module, an updated Historical Temporal Shipping (HTS) database containing information on commercial ships and fishing vessels with a simulated vessel movement module;
- Dynamic Ambient Noise Module (DANM), successor to the ANDES program; and
- Reported Vessel Module.

DAPS was designed to predict the azimuthal dependence of ocean noise in the 25-5,000-Hz frequency band, including surface shipping and wind-generated noise. Lloyds of London records were used for initial shipping spatial distributions, and ship tracks were inferred from shipping lanes as input to a propagation model. Fishing vessel activity used historical vessel distributions and fishery statistics collected by the Food and Agricultural Organization of the United Nations to incorporate fishing vessel densities. The wind-generated component was obtained from the Surface Marine Gridded Climatology and empirical relations between the ocean ambient noise levels and wind speed. The DANM module presently is being reviewed by the U.S. Navy's Oceanographic and Atmospheric Master Library (OAML), which is responsible for maintaining and distributing standardized databases and models to the U.S. Navy fleet. If successful, DANM will be the first ambient noise model to obtain OAML approval.

MODELING THE EFFECTS OF NOISE ON MARINE MAMMALS

A conceptual model can assist in describing the interactions necessary to assess the impact of ocean noise on marine mammals and other marine animals. The ocean noise input to the system of marine mammals consists of all types of ocean noise, including those generated naturally by physical and biological means and those generated from human activities (Chapter 2). The system being evaluated consists of marine mammals and in the simplest terms can be treated as multiple environmental and physical factors on which the ocean acts to produce the output. The output consists of metrics that can be used to assess the impact of the ocean noise on the

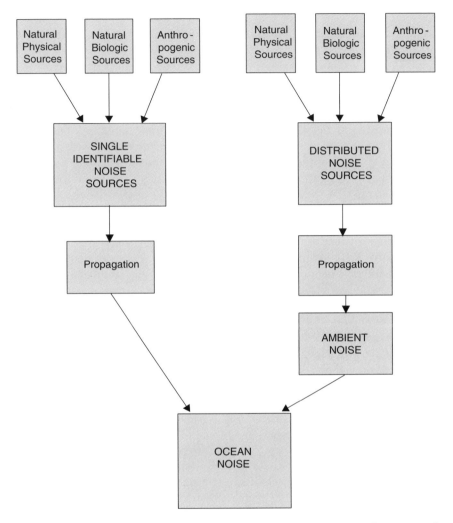

FIGURE 4-3   Components of the ocean noise input to the overall conceptual model.

system.   Such measures may be physiologically based, such as noise levels that produce temporary or permanent threshold shifts in given animals, or behaviorally based, such as sound levels that cause cessation of mating calls.   Ocean noise can be dispersed (Figure 4-3) and is capable of incorporating available (sub)models.   These existing models can be used to determine the appropriate input to a model and evaluate a given scenario.

A model of effects that predicts the impact of acoustic signals on marine mammals should consist of six main components: (1) a description of the source time function or source spectrum and level and a model of the source distribution, (2) physical oceanographic and geoacoustic databases, (3) models of marine mammals distribution in three dimensions to determine exposure, (4) models to predict the sound signal at an animal, (5) biological databases and models for marine mammal hearing and movement, and (6) population-based models to look for effects at these levels.

Recent breakthroughs in the understanding of the effects of noise on animal hearing along with developments in the understanding of acoustic propagation have enabled the combination of hearing models with acoustic models, referred to as integrative models. These integrative models include a physical oceanographic component that controls the propagation of sound. Integrated models include a library of common sound sources (biological and man-made), environmental features that affect sound propagation such as bathymetry and ocean dynamics, and algorithms for modeling sound propagation, such as PE models. Biological components are divided into the following: animal distribution databases; animal behavior data and models, including migration, diving patterns, and behavioral responses to sounds; and models for the mechanical and neural responses to sound by the organism. Systems architecture of integrative models can be designed to include data synthesis display and communications tools that enable investigators to work as a distributed network and databases and modeling algorithms that are shared among widely distributed universities, labs, and data centers. The goals of these models or their successors are to predict the outcome of a given sound exposure regimen and to represent that information in dynamic graphical displays and probabilistic functions. In other words, model predictions will be quantitative with quantified limits of uncertainty.

One example of these new and integrative animal effects models is the ESME (Effects of Sound on the Marine Environment) model, sponsored by ONR. At present ESME is halfway through the four-year development plan. ESME identifies the elements necessary for a predictive risk assessment model and develops an architecture for fitting the pieces together. Not all of the necessary databases are full, and gaps in understanding still exist. To date, a basic structure has been developed and applied to two simple but realistic test problems. The first scenario dealt with the effects of noise on dolphins in the Southern California Bight. The second test problem focused on dolphins in the Middle Atlantic Bight south of Rhode Island. These test problems assess intermodule communications and test different databases and modeling algorithms. The tests also examine different configurations for ESME and its successors. Several alternative models for further development will likely result from ESME, ranging from the simple to the most complex.

A relatively simple integrative model, such as the Acoustic Integration Model (AIM), could be PC based to enable a wider range of users to experiment with underwater sound scenarios, providing educational, scientific, and environmental management functions. AIM was designed to model the movements and behaviors of acoustic sources and receivers. These receivers are virtual animals and have been dubbed animats. The AIM model interfaces with another acoustic propagation model that simulates the acoustic field produced by the acoustic source(s). The animats can be programmed to simulate natural responses, including reactions to the sound field. The acoustic history of each animat is recorded, a valuable and important output. The model allows multiple Monte Carlo model simulations to estimate the impact of various scenarios. At the other end of the scale, ESME or an equivalent tool might be integrated with state-of-the-art complex multidimensional physical ocean models running on supercomputers. A complex version of the integrative tools would clearly limit accessibility to only the most sophisticated users but would offer the greatest possible flexibility and accuracy.

At present, the integrative models are concentrating on modeling effects of individual sound sources on individual animals or individuals within pods. The effects of distributed sources are not, at this time, being investigated. Hearing at the individual level is being modeled at several levels, from the micromechanical activity of the inner ear through whole head resonance. Inner ear models are based on basilar membrane response data for well-studied ears, especially in vivo measurements in mice, cats, and gerbils. Inner ear structural data on these ears are being compared with parallel data from representative marine mammal ears (mysticete, odontocete, and pinniped) in order to modify the inner ear response models to accurately represent stiffness and mass variations in marine mammals compared to smaller land mammals. This will affect both sensitivity and frequency responses in marine ears. Middle ear and whole head responses, particularly head transfer functions, are two areas for which no adequate land analog exists. Models for these elements of hearing are being formed based on direct measures of marine mammal tissue mechanical characteristics, acoustic impedances, and complex tissue resonance.

No attempts have been made to model the effects of noise on the habitat and ecosystem of marine mammals. Fish and other marine organisms respond to noise in both experimental systems and marine environments. Because they are prey items for some marine mammals and are important components of the marine ecosystem, it is also necessary to examine the effects of noise on these organisms to incorporate all of the effects of noise on marine mammals.

DATABASES

Ancillary Data for Effects Modeling

The data necessary to allow modeling of the overall effect of ocean noise on marine mammals are quite varied and in general do not yet exist in the volume and completeness needed. There are three major categories of data that are required: (1) data that characterize sources, (2) data that characterize how acoustic energy propagates from the source to the animal, and (3) data that characterize the effects that sounds have on marine mammals and fishes, both physiological and behavioral.

The information needed about the sources is the characterization of the source itself, such as its output level, its frequency band, and so on, as discussed at the beginning of this chapter, and the activity level of the sources, where they are operating, and when. One needs to know the velocities, densities, and attenuation factors in the water column and in the upper strata below the seafloor to describe accurately the propagation of sound waves from the source to the animal some distance away. The information is needed to characterize the effects that sounds have on marine mammals and the specific research topic being examined.

Data do exist that fall into all three of these categories, but they are incomplete, scattered, and, in many cases, inaccessible for national security reasons. However, two programs are specifically addressing the potential impact of ocean acoustic noise on the marine environment by developing comprehensive databases. These databases presently will be used for rapid data retrieval, mapping, and statistical correlation studies, but they also could be used as inputs to future physics-based, numerical modeling efforts. One is the Sound, Oceanography, and Living Marine Resources (SOLMAR) program at the NATO Supreme Allied Commander Atlantic Undersea Research Centre in La Spezia, Italy. Data, primarily from the Mediterranean Sea, are being assembled on the occurrence of cetacean strandings, results of visual surveys, and underwater acoustic recordings of vocalizations. Standard oceanographic and geophysical measurements such as water mineral and chlorophyll content, conductivity/temperature/depth profiles, and bathymetry data are also being collected. In addition, satellite-based measurements such as altimetry, sea color, and sea surface temperature are also being collected. SOLMAR databases are applied to a Geographical Information System (GIS) framework. Another program, the Living Marine Resources Information System, is developing databases of global distributions of marine animal species with no acoustic data. The primary sources of information presently are the National Marine Fisheries Service visual survey reports, as well as other publications in the open literature; however these data are confined largely to coastal areas.

The Census of Marine Life is a new international effort to determine numbers and types of marine organisms and their geographic and depth distribution worldwide. One part of this program is an open-access, Internet-based collection of databases and associated processing tools called the Ocean Biogeographic Information System (OBIS). The databases will initially include components such as a history of marine animal populations, biogeoinformatics of Hexacorallia for corals and sea anemones, and data on the chemosynthetic ecosystems in the Arctic and northern Atlantic Oceans. Also included are oceanographic and environmental databases, all in a GIS framework to permit full ecological system assessment. The computer and communications-based setting is expected to permit computational functionality among internationally distributed systems.

Organized information about ports and shipping lanes is maintained by the U.S. Navy's Space and Naval Warfare Systems Command, which defines 521 ports and 3,762 traffic lanes. Lloyds of London maintains information about the merchant fleets of the world, the number of ships in each ship-type category, and gross tonnage.

Oil industry activity that contributes most to ocean noise can be monitored by subscribing to any of a number of commercial information services. For example, IHS Energy provides relatively comprehensive information dating back to 1994 about individual marine seismic crews and where they are, and have been, working. The location data may be no more specific than "North Sea," and no information is given about the specifications of the air-gun arrays being used. IHS Energy will research its database and generate reports for a fee. A similar type of service is provided by ODS-Petrodata with regard to offshore mobile and platform drilling rigs. Together, these two services can supply an overall picture of where these noise sources are in time and space, but neither provides information about the noise generated by these operations. Measurements need to be recorded of different drilling techniques in different environments to determine if they make enough noise to cause concern. If so, then a catalogue of the noise output of the different techniques should be maintained and used to calculate the contribution to the noise budget from drilling rigs. Because the information is considered proprietary, it is unlikely that the details of air-gun arrays will be included in the seismic crew databases. Using published values of air-gun array source levels of 260 dB re 1 μPa-m, peak-to-peak (Richardson et al., 1995), will produce estimates that err safely on the high side. This level is best used for the output oriented vertically, and for the horizontally oriented output, a number around 235 dB re 1 μPa-m, peak-to-peak, is more suitable.

There appears to be no suitable, all-inclusive source of information about offshore construction activities outside of the oil and gas industry. These activities include cable laying, dredging and reclamation projects, tunnel boring and bridge building, dock building, and port construction. Petrodata's Marine and Coastal Construction System is a database and newswire service that provides information on planned worldwide marine and coastal construction projects. It seems most adept at capturing projects located in Europe, Asia, and the Middle East. Again, measurements of the noise created by these activities are not numerous, so this is an area where much work needs to be done if an assessment is to be made as to the importance of these activities to the ocean noise budget.

Data on the physical properties of the ocean waters and the near-seafloor sediments exist in detail in some places and are nonexistent in others. It is beyond the scope of this report to discuss this topic here other than to say that in some instances it is crucial to know the details of the seafloor topography, the details of the water column sound speed and absorption properties, and the details of the seismic velocities, densities, and absorption properties of the strata below the seafloor.

Although there is an extensive literature on the effects of sound on marine mammals, it is patchy and inconclusive. A tremendous amount of work remains to be done to determine the effects of sound on marine mammals. In particular there have been few studies to relate specific dosage of sound to effects likely to be of biological significance. One of the recommendations of this report is that a single federal agency or organization be charged with the responsibility of overseeing all of these activities, including all data collection. As more and more locations around the world place restrictions on activities that create noise in ocean waters, and such restrictions cause data to be collected with regard to this issue, it seems prudent to establish an official body that catalogues these different data sets, if it does not actually oversee the storage and archiving of them.

Ocean Noise Databases

Currently there is no coordinated program to organize, support, and execute an ongoing data collection effort to supplement the general ambient noise data sets that were the basis of empirical curves such as those of Wenz and Knudsen. There are ongoing individual efforts, but they are incomplete, scattered, and in some cases may not be available because of national security reasons within the United States and other nations. Typically, these efforts are focused on averaged values of the acoustic pressure spectrum and transients are excluded. One significant collection is the archived information of the U.S. Navy, held by the Naval Oceanographic Office (NAVOCEANO). Nearly 50,000 omnidirectional measurements of ambient marine noise are held within the NAVOCEANO Data Warehouse. Data collection began in the 1950s and is organized by season, frequency, location, and time. NAVOCEANO also maintains wind noise estimates based on the model projections using adaptations of the Wenz curves (Plate 6a-d).

Representative samples of NAVOCEANO archives for two seasons (summer and winter) and two frequencies (50 and 3,500 Hz) highlight the potential usefulness of such a dataset (Plate 7a-d). The data collected are oriented to geographic regions of past, current, and future naval operations interests. Data measurements vary in duration of collection, from very short (<1 hour) to drifting buoys that gather data for weeks to months. Through careful analysis data collected in the presence of known contaminants (seismic sources, nearby passing ships) were discarded. Perhaps the most striking feature of these figures is the lack of data in most of the world's oceans.

Additional noise databases can be found in Etter (1996). Etter's Table 10.3 lists the Advanced Environmental Acoustic Support data bank as well as the NAVOCEANO database, and Table 10.5 contains noise databases that reside in the OAML. These OAML-approved databases include three shipping noise databases that cover all of the northern hemisphere as a function of season, Arctic noise near the marginal ice zone on a monthly basis, and the wind and residual noise database, which provides monthly variations in noise levels not containing shipping for the northern hemisphere.

Access to the databases listed is restricted, making it difficult to review them and use them for scientific purposes. All were gathered to meet U.S. Navy sonar system needs. Much of the data probably were not collected in a systematic way using fixed procedures. A clear bias toward the horthern hemisphere exists.

Other ambient noise data sets can be found in various places outside the operational navy community. As one example, the National Oceanographic and Atmospheric Administration has been collecting SOSUS (Sound Surveillance System) data off the coast of the State of Washington since 1991 (Chris Fox, personal communication to committee, 2001). However, at present, a major gap in existing noise databases is that no long-term (greater than a decade), systematically collected, ocean acoustic data set exists for any frequency band.

Additional gaps in marine noise databases include the facts that no noise database is known to exist for the southern hemisphere except the set of measurements made around the continent of Australia by the Defence Science and Technology Office, and possibly those in the waters off New Zealand. In addition, no systematic noise monitoring data set has been collected in biologically sensitive areas for specific species. Finally, if the whole frequency band from 1 Hz to 200 kHz is taken as the band of interest, a gap exists in databases at frequencies above several kilohertz. Additional planning is required in collecting data at high frequencies because of the large data rates involved; continuous sampling is not practical unless some type of real-time processing is implemented.

A well-recognized issue with ambient noise measurements, particularly in shallow water, is the effect of the propagation characteristics on the received field. Therefore, a gap in existing shallow-water noise databases is lack of knowledge of the ocean-bottom geoacoustic properties in the regions where the measurements were made. More generally, the quality of ambient noise databases is directly related to the quality and variety of ancillary information (e.g., near-surface winds, shipping traffic, visual observations of marine animals) collected simultaneously at the same location. Development of a long-term ocean noise monitoring system requires careful consideration of which types and in what ways this supporting information will be collected.

SUMMARY

Sound sources in the ocean can be categorized and modeled as two main types: unknown distributed sources (that is, unknown location, source level, and spectral content) referred to as ambient noise and best modeled as statistical in nature, and identified single sources best modeled deterministically. Noise from the collection from all sources is referred to as "ocean noise" in this report. The dominant source of ambient noise is associated with ocean surface wave activity. In the frequency band from 5 to 200 Hz, shipping may be dominant, at least in the northern hemisphere. The time-averaged received levels of shipping noise in some locations can be fairly well modeled. Above 200 Hz, noise levels from breaking waves are roughly modeled through the use of empirical relations between noise level and wind speed. Limitations exist in ambient noise models not just from lack of knowledge of the source characteristics and distributions but also resulting from uncertainties in the environment. The sounds from single sources, such as sonar and air-guns, are usually well modeled by propagation codes. The accuracy of these models is limited by environmental uncertainty. The effects of sound from single sources on marine mammals are beginning to be modeled by integrative tools such as AIM and ESME. The effects of distributed sources, such as shipping and wind, on marine mammals are not yet well modeled.

From field observations and threshold experiments on captive animals (see Chapter 3), it is clear that sound can disturb marine mammals both behaviorally and physiologically. Noise from shipping may be affecting marine mammals adversely. Similarly, high-intensity transient sources at short ranges may have significant effects on marine mammal physiology or behavior. Modeling these effects is possible and prudent. While modeling the physiologic effects is relatively straightforward, modeling behavioral effects is difficult and needs more effort. In all cases, field data must be collected to validate the model predictions.

5

Findings and Recommendations

This report is the third in a series by the National Research Council examining the potential effects of ocean noise on marine mammals. Although these reports evolved from very different charges and were generated by separate committees, similar research needs became evident during each study. The recommendations in this report expand on rather than replace those from earlier efforts (Appendix D; NRC, 1994, 2000). Recommendations of all three reports should be examined to better comprehend the full spectrum of research required to understand the effects of human-generated noise on the marine ecosystem. It should also be noted that while some of the research needs from past reports, particularly from the first report (NRC, 1994), have been met, some of the new information has led to additional research questions that must now be answered.

SOURCES OF NOISE IN THE MARINE ENVIRONMENT

The recommendations made here are intended to improve our understanding of the effects of noise on marine mammals. To this end, any efforts to implement these recommendations should be planned and structured to facilitate use in conjunction with data on marine mammal physiology and behavior.

Currently, data regarding noise produced by shipping, seismic surveying, oil and gas production, marine and coastal construction, and other marine activities are either not known or are difficult to analyze because they are maintained by separate organizations such as industry database companies, shipping industry groups, and military organizations. It would be advantageous to have all data in a single database in order to improve the ability of interested parties to access the data sets and use them in research, for scientific publications, in education, and for management and regulatory purposes. This database could be a distributed network of linked databases, using a standardized series of units of measure. International cooperation in this database development effort, as well as international access to the information, should be encouraged since the marine mammal and ocean noise issue is global.

Recommendation: *Existing data on marine noise from anthropogenic sources should be collected, centralized, organized, and analyzed to provide a reference database, to establish the limitations of research to date, and to better understand noise in the ocean.*

Each characteristic of noise from anthropogenic sources may differentially impact each species of marine mammals. The complex interactions of sound with marine life are not sufficiently understood to specify which features of the acoustic signal are important for specific impacts. Therefore as many as characteristics as possible should be measured and reported.

Recommendation: *Acoustic signal characteristics of anthropogenic sources (such as frequency content, rise time, pressure and particle velocity time series, zero-to-peak and peak-to-peak amplitude, mean squared amplitude, duration, integral of mean squared amplitude over duration, repetition rate) should be fully reported. For transients, publication of actual acoustic pressure time series would be useful. Experiments should be conducted that expose marine mammals to variations in these characteristics in order to determine the physiological and behavioral responses to different characteristics. Particular attention should be paid to the sources that are likely to be the large contributors to ocean noise in particularly significant geographical areas and to sources suspected of having significant impacts on marine life.*

Little is known about long-term trends in ocean noise levels. Although evidence is limited concerning long-term trends in ocean noise, and few observations concerning the effects of ocean noise on marine life exist, the current data are sufficient to warrant increased research and attention to trends in ocean noise.

Recommendation: *A long-term ocean noise monitoring program over a broad frequency range (1 Hz to 200 kHz) should be initiated. Monitoring and data analysis should include average or steady-state ambient noise as well as identifiable sounds such as seismic surveying sources, sonars, and explosive noise that are not identified in classical ambient noise data sets. Acoustic data collection should be incorporated into global ocean observing systems initiated and under discussion in the United States and elsewhere. A research program should be initiated that develops a predictive* *model of long-term noise trends. Data from monitoring systems should be available in a timely manner to facilitate informed decision making by interested industry, military, and marine researchers, operators, and regulatory agencies.*

Efforts must be made to measure ocean noise in marine mammal habitats. Until these habitats are fully known and described, it is reasonable to begin a long-term monitoring program in coastal areas and areas close to known marine mammal foraging, migration, and breeding areas.

Recommendation: *Efforts to measure ocean noise should be targeted toward important marine mammal habitats. As new marine mammal habitats are identified, these should be added to the acoustic surveys in order to provide a complete picture of the acoustic environment in important marine mammal ecosystems.*

Identifying reliable indicators for anthropogenic sources will provide an additional modeling tool and predictive capability that will be particularly useful in areas where long-term monitoring may be difficult or impossible. For instance, although the global shipping fleet increased from 30,000 commercial vessels in 1950 to 87,000 vessels in 1998, consequent noise changes cannot be determined because noise data were not collected in a systematic way to allow scientific comparisons, nor are they being systematically collected at this time. Similar needs exist for every facet of human activity in the oceans.

Recommendation: *Research to determine quantitative relationships between levels of anthropogenic activity and noise should be conducted. For example, if there is a robust relationship between vessel type and noise, vessel traffic data could be used to predict shipping noise.*

MARINE MAMMALS AND OCEAN NOISE

Although it is difficult to obtain direct evidence of impacts of human activity on marine mammals, it is even more difficult to determine long-term impacts on individuals or impacts on populations. Although the few documented cases of direct impact on individuals have raised awareness of potential population impacts, no measures exist of marine mammal population effects from ocean noise.

Recommendation: *Whenever possible, all research conducted on marine mammals should be structured to allow predictions of whether responses observed indicate population-level effects.*

Despite the large body of marine mammal research to date, including what was recommended in previous reports (NRC, 1994), there is a surprising lack of information regarding the global distribution of marine mammals. Migration routes, breeding grounds, and feeding areas are known for relatively few species. In order to predict the importance of noise effects on marine mammal behavior, the seasonal and geographic distribution of the mammals must be better known both through survey data and through the use of predictive oceanographic variables, such as topography, bottom type, and water column variables. This enormous task will require the development of new sampling and extrapolation techniques in order to be practically achievable.

Recommendation: *Research should be conducted beyond locales already known and studied to globally characterize marine mammal distributions and populations.*

While good progress has been made in describing marine mammal acoustic repertoires, much less is known about the details of natural patterns of sound production, including the means of production and context in which different vocalizations are produced, as well as how they vary diurnally, seasonally, and geographically. Marine mammals themselves may be significant sources of ocean noise, although possibly in localized areas over limited time periods.

Recommendation: *Research should be undertaken to describe the distribution and characteristics of sounds generated by marine mammals and other marine organisms seasonally, geographically, and within behavioral contexts. These studies will also shed light on the contribution which marine organisms make to the global ocean noise budget.*

Efforts to improve marine mammal tagging technology should continue to receive support. Two technological improvements of current tags are needed: (1) increase the duration of long-term data-gathering tags from months to multiple years to observe annual behavior cycles and migration patterns, and (2) extend the duration of high-resolution tags from hours to days to gather more data on daily behavior and environmental cues. Current tagging technology allows individual marine mammals to be tracked up to months. Tags capable of higher-resolution data collection, including animal orientation, acceleration, and produced or received sounds, can generally collect data for less than one day. These data have proven very valuable in determining behavioral patterns in a variety of cetaceans and pinnipeds and correlating their behavior with environmental cues. The technology should continue to be developed to allow longer studies using both the high- and low-resolution tags.

Recommendation: *Marine mammal tagging studies should be continued to observe behavioral changes in response to acoustic cues and to provide important data for simulation models.*

Short-term responses of marine mammals to anthropogenic noise sources have been documented to a limited degree; however, long-term effects of marine noise on the behavior of marine mammals have received less attention. Impacts due to increases in background ambient noise have not been documented.

Recommendation: *Research should be conducted to determine subtle changes in marine mammal behavior, as well as failure to detect calls from other animals or echoes from their own echolocation, that might result from masking of biologically important acoustic information by anthropogenic sounds.*

Stress indicators may be one useful marker for long-term effects of anthropogenic noise on marine mammals.

Recommendation: *Research efforts should seek to determine if reliable long-term stress indicators exist and if they can be used to differentiate between noise-induced stress and other sources of stress in representative marine mammal species.*

Fish use sound in many ways that are comparable to the ways marine mammals communicate and sense their environment. The effects of anthropogenic noise on fishes and other nonmammalian species, including their eggs and larvae, are largely unknown. As cohabitants of the marine ecosystem and as members of the same food web, noise impacts on marine fish could, in turn, affect marine mammals.

Recommendation: *The impact of noise on nonmammalian organisms in the marine ecosystem should be examined.*

OCEAN NOISE MODELS

Simulation models that predict the characteristics of the noise (frequency content, mean squared level, peak level, pressure time series, etc.) and their effects on marine mammals may assist in understanding and mitigating harmful effects of marine noise on mammals. At least one such effort is underway: the Effects of Sound on the Marine Environment model sponsored by the Office of Naval Research. Modeling some direct physiologic effects on hearing (e.g., temporary or permanent threshold shift) is relatively straightforward, although limited by the small data sets available from a limited number of species. These integrative tools should be expanded to include the effects of sources of noise that may change their distribution over time such as shipping, wind-induced breaking waves, and distributed biologic noise. More effort should be placed on modeling, both explicit marine species hearing models and behavioral effects models for all types of ocean noise.

Recommendation: *Modeling efforts that integrate acoustic sources,* *propagation, and marine mammals should be continued and fully supported.*

The conventional approach that utilizes an average pressure spectrum budget is limited in its application to the marine mammal problem. A more comprehensive approach that encompasses contributions of both transient events and continuous sources to ocean noise should be pursued. Many of this committee's recommendations, particularly those concerning information on distribution and source signatures of man-made sources, must be addressed in order to have the capability to develop a marine-mammal-relevant global ocean noise model. In addition, since model validation is a critical part of the model development process, the committee's recommendations pertaining to the collection of high-quality, well-documented ocean noise data sets must be pursued in tandem.

Recommendation: *A model of global ocean noise that properly reflects the impact of both ambient noise and noise from identified sources on marine mammals should be developed and verified.*

OVERARCHING RECOMMENDATIONS

Federal leadership is needed to (1) monitor ocean noise, especially in areas with resident marine mammal populations; (2) collect and analyze existing databases of marine activity; and (3) coordinate research efforts to determine long-term trends in marine noise and the possible consequences for marine life.

Recommendation: *A federal agency should be mandated to investigate and monitor marine noise and the possible long-term effects on marine life by serving as a sponsor for research on ocean noise, the effects of noise on marine mammals, and long-term trends in ocean noise.*

Recent reports both in the press and from federal and scientific sources indicate that there is an association between the use of high-energy mid-range sonars and some mass strandings of beaked whales. Recent mass strandings of beaked whales have occurred in close association, both in terms of timing and location, with military exercises employing multiple high-energy, mid-frequency (1-10 kHz) sonars. In addition, a review of earlier beaked whale strandings further reinforced the expectation that there is at least an indirect relationship between strandings and the use of multiple mid-range sonars in military exercises in some nearshore beaked whale habitats. Several press reports about the recent incidents appeared while this report was in preparation attributing the strandings to "acoustic trauma." Acoustic trauma is a very explicit form of injury. In the beaked whale cases to date, the traumas that were observed can result from many causes, both directly and indirectly associated with sound, but similar traumas have been observed in terrestrial mammals under circumstances having no relation to sound exposure. Careful sampling and analysis of whole animals have rarely been possible in the beaked whale cases to date, which has made definitive diagnoses problematic. To date, eight specimens in relatively fresh condition have been rigorously analyzed. Because of the repeated associations in time and location of the strandings and sonar in military exercises, the correlation between sonars and the strandings is compelling, but that association is not synonymous with a causal mechanism for the deaths of the stranded animals. The cause of death in all cases was attributed to hyperthermia, but a precise cause for the unusual traumas that were also seen in the cases examined has not yet been determined. The NATO/SACLANT Undersea Research Center report (D'Amico and Verboom, 1998) and the joint NOAA-Navy interim report (Evans and England, 2001) have not been discussed in detail in this document because of the preliminary nature of the findings. However, this is clearly a subject to which much additional research needs to be directed.

Recommendation: *A program should be instituted to investigate carefully the causal mechanisms that may explain the traumas observed in beaked whales, whether this is a species-specific or broader issue, and how the acoustics of high-energy, mid-range sonars may directly or indirectly relate to mass stranding events. The research program outlined in Evans and England (2001) represents a good initial effort.*

Addressing the challenge of both short- and long-term effects of ocean noise on marine mammals is a difficult problem and will require a multidisciplinary effort between biologists and acousticians to establish a rigorous observational, theoretical, and modeling program. An initial significant focus of this work should be the examination of the possible relationship between the acoustics of identifiable high-energy, mid-frequency sonars, marine mammal trauma, and mass stranding events. In addition, a study of the potential influence of ambient noise on long-term animal behavior should be vigorously pursued.

Recommendation: *A research program should be instituted to investigate the possible causal relationships between the ambient and identifiable source components of ocean noise and their short- and long-term effects on marine organisms.*

The public, including environmental advocates, are very interested in anthropogenic noise in the ocean and its effect on marine animals. Recently there has been a communication gap between users of sound in the ocean, including scientists, and the public. Much of the gap in understanding between the ocean science community and the public arises from the public's lack of understanding of fundamental acoustic concepts and the scientific community's failure to communicate those concepts effectively. Source and received levels, propagation loss, air-water physical acoustic differences, and the term "decibel" are examples of concepts that have been misunderstood by the media, environmental organizations, and the general public.

Recommendation: *The committee encourages the acoustical oceanography community, marine mammal biologists, marine bioacousticians, and other users of sound in the ocean, such as the military and oil industry, to make greater efforts to raise public awareness of fundamental acoustic concepts in marine biology and ocean science so that they are better able to understand the problems, the need for research, and the considerable potential for solving noise problems.*

References

Aki, K., and P.G. Richards. 1980. Quantitative Seismology—Theory and Methods. W. H. Freeman, San Francisco, CA, pp. 533-534.

American National Standard Institute (ANSI). 1994. Acoustical Terminology. Document Number: ANSI/ASA S1.1-1994(R1999) 01-Jan-1994. American National Standards of the Acoustical Society of America, 52 pp.

Andrew, R.K., B.M. Howe, J.A. Mercer, and M.A. Dzieciuch. 2002. Ocean ambient sound: Comparing the 1960s with the 1990s for a receiver off the California coast. Acoustics Research Letters Online 3(2):65-70.

Applied Acoustics. 1977. Ambient sea noise dependence on local, regional and geostrophic wind speeds: Implications for forecasting noise. Tavener S. Institution. Defense Science and Technology Organization. Pyrmont, NSW, Australia. Applied Acoustics 51(3):317-338.

Arase, E.M., and T. Arase. 1967. Deep-sea ambient-noise statistics. Journal of the Acoustical Society of America 44:1679-1684.

Au, W. 1993. The sonar of dolphins. Springer-Verlag, New York, 277 pp.

Au, W.W.L., and K. Banks. 1998. The acoustics of the snapping shrimp (Synalpheus parneomeris) in Kaneohe Bay. Journal of the Acoustical Society of America 103:41-47.

Au, W.W.L., and M. Green. 2000. Acoustic interaction of humpback whales and whale-watching boats. Marine Environmental Research 49:469-481.

Au, W.W.L., and P.W.B. Moore. 1984. Receiving beam patterns and directivity incidents of the Atlantic bottlenose dolphin. Journal of the Acoustical Society of America 75:255-262.

Au, W.W.L., D.A. Carder, R.H. Penner, and B.L. Scronce. 1985. Demonstration of adaptation in beluga whale echolocation signals. Journal of the Acoustical Society of America 77:726-730.

Au, W.W.L., J. Mobley, W.C. Burgess, M.O. Lammers, and P.E. Nachtigall. 2000a. Seasonal and diurnal trends of chorusing humpback whales wintering in waters off western Maui. Marine Mammal Science 16:530-544.

Au, W.W.L., A.N. Popper, and R.R. Fay, eds. 2000b. Hearing by whales and dolphins. Springer-Verlag, New York, 485 pp.

Bacon, S., and D.J.T. Carter. 1993. A connection between mean wave height and atmospheric pressure gradient in the North Atlantic. International Journal of Climatology 13:423-436.

Baggeroer, A., and W. Munk. 1992. The Heard Island Feasibility Test. Physics Today 45:22-30.

Bain, D.E., and M.E. Dahlheim. 1994. Effects of masking noise on detection thresholds of killer whales. Pp. 243-256 in Marine Mammals and the Exxon Valdez. T.R. Loughlin, ed. 4th Edition. Academic Press, San Diego.

Baird, I.L. 1974. Some aspects of the comparative anatomy and evolution of he inner ear in submammalian vertebrates. Brain Behavior and Evolution 10:11-36.

Bannister, B.R., D.R. Melton, and G.E. Taylor. 1989. Testability of digital circuits via the spectral domain. Pp. 340-343 in IEEE International Conference on Computer Design.

Barlett, M.L., and G.R. Wilson. 2002. Characteristics of small boat signatures. Journal of the Acoustical Society of America 112(part 2):2221.

Bartol, S.M., J.A. Musick, and M. Lenhardt. 1999. Auditory evoked potentials of the loggerhead sea turtle (Caretta caretta). Copeia 99(3):836-840.

Bauer, G.B., J.R. Mobley, and L.M. Herman. 1993. Responses of wintering humpback whales to vessel traffic. Journal of the Acoustical Society of America 94:1848.

Bendat, J.S., and A.G. Piersol. 1986. Random Data: Analysis and Measurement Procedures. 2nd Edition. John Wiley and Sons, New York.

Blane, J.M., and R. Jaakson. 1994. The impact of ecotourism boats on the St. Lawrence beluga whales. Environmental Conservation 21:267-269.

Borggaard, D., J. Lien, and P. Stevick. 1999. Assessing the effects of industrial activity on large cetaceans in Trinity Bay, Newfoundland. Aquatic Mammals 25:149-161.

Bowles, A.E., M. Smultea, B. Würsig, D.P. DeMaster, and D. Palka. 1994. Relative abundance and behavior of marine mammals exposed to transmissions from the Heard Island Feasibility Test. Journal of the Acoustical Society of America 96:2469-2484.

Breeding, J.E., Jr. 1993. Description of a noise model for shallow water: RANDI-III. Journal of the Acoustical Society of America 94(part 2):1920.

Bregman, A.S. 1990. Auditory Scene Analysis. MIT Press, Cambridge, MA, 773 pp.

Brekhovskikh, L.M., and Y. Lysanov. 1991. Fundamentals of Ocean Acoustics. 2nd Edition. Springer-Verlag, New York, 270 pp.

Bryant, P.J., C.M. Lafferty, and S.K. Lafferty. 1984. Reoccupation of Laguna Guerrero Negro, Baja California, Mexico, by gray whales. Pp. 375-386 in The Gray Whale Eschrichtius robustus, M. L. Jones et al., eds. Academic Press, Orlando, FL.

Buck, B.M., and J.H. Wilson. 1986. Nearfield noise measurements from an Arctic pressure ridge. Journal of the Acoustical Society of America 80:256-264.

Buckingham, M.J., and J. R. Potter, eds. 1995. Sea Surface Sound '94. Proceedings of the 1994 Lake Arrowhead Conference. World Scientific Publishing Co., 494 pp.

Budelmann, B.U. 1988. Morphological diversity of equilibrium receptor systems in aquatic invertebrates. Pp. 757-782 in Sensory Biology of Aquatic Animals, J. Atema et al., eds. Springer-Verlag, New York.

Budelmann, B.U. 1992. Hearing in crustacea. Pp. 131-139 in Evolutionary Biology of Hearing, D.B. Webster et al., eds. Springer-Verlag, New York.

Burdic, W.S. 1991. Underwater Acoustic System Analysis. 2nd Edition. Prentice-Hall Englewood Cliffs, NJ, 466 pp.

Burgess, W.C. 2001. The bioacoustic probe: A miniature acoustic recording tag. Abstract. Presented to 14th Biennial Conference on the Biology of Marine Mammals, November/ December. Vancouver, BC, Canada.

Burgess, W.C., P.L. Tyack, B.J. Le Boeuf, and D.P. Costa. 1998. A programmable acoustic recording tag and first results from free-ranging northern elephant seals. Deep-Sea Research Part II 45:1327-1351.

Burns, J.J., and G.A. Seaman. 1985. Investigations of belukha whales in coastal waters of western and northern Alaska II—Biology and ecology. U.S. National Oceanic and Atmospheric Administration, 129 pp.

Busnel, R.G., ed. 1963. Acoustic Behavior of Animals. Elsevier, Amsterdam.

Caldwell, M.C., and D.K. Caldwell. 1965. Individualized whistle contours in bottlenosed dolphins (*Tursiops truncatus*). Nature 207:434-435.

Calkins, D.G. 1979. Marine mammals of Lower Cook Inlet and the potential for impact from outer continental shelf oil and gas exploration, development, and transport. Pp. 171-263 in Environmental Assessment of the Alaskan Continental Shelf: Final Reports of Principal Investigators, Volume 20. U.S. National Oceanic and Atmospheric Administration, Juneau, AK.

Cato, D.H. 1978. Marine biological choruses observed in tropical waters near Australia. Journal of the Acoustical Society of America 64:736-743.

Cato, D.H. 1980. Some unusual sounds of apparent biological origin responsible for sustained background noise in the Timor Sea. Journal of the Acoustical Society of America 68:1056-1060.

Cato, D.H. 1997a. Ambient sea noise in Australian waters. Proceedings of the 5th International Congress on Sound and Vibration, International Institute of Acoustics and Vibration, p. 2813.

Cato, D.H. 1997b. Features of ambient noise in shallow water. Proceedings of International Conference on Shallow-Water Acoustics (SWAC'97), pp. 385-390.

Cato, D. 1992. The biological contribution to the ambient noise in waters near Australia. Acoustics Australia 20:76-80.

Cato, D.H. 2001. *Doug Cato Notes.* Marine Mammals Comparison of Natural Ambient Noise with Traffic Noise, Woods Hole, MA, June.

Cato, D.H., and M.J. Bell. 1992. Ultrasonic ambient noise in Australian shallow waters at frequencies up to 200 kHz. DSTO Materials Research Laboratory Report No. MRL-TR-91-23. Defense Science and Technology Organization, Ascot Vale, Victoria, Australia.

Cato, D.H., and R.D. McCauley. 2002. Australian research in ambient sea noise. Acoustics Australia 30:13-20.

Cavanagh, R.C. 1974a. Fast Ambient Noise Model I (FANM I). Acoustic Environmental Support Detachment, Office of Naval Research.

Cavanagh, R.C. 1974b. Fast Ambient Noise Model II (FANM II). Acoustic Environmental Support Detachment, Office of Naval Research.

Cavanagh, R.C. 1978. Acoustic fluctuation modeling and system performance estimation, Volume I. Science Applications, Inc.

Chapman, C.J., and A.D. Hawkins. 1973. A field study of hearing in the cod (*Gadus morhua L.*). Journal of Comparative Physiology A 85:147-167.

Copeland, G.J.M. 1993. Low frequency ambient noise—generalized spectra in natural physical sources of underwater sound. Pp. 17-30 in Sea Surface Sound, B.R. Kerman, ed. Kluwer Academic Publishing, Dordrecht, Holland.

Cosens, S.E., and L.P. Dueck. 1988. Responses of migrating narwhal and beluga to icebreaker traffic at the Admiralty Inlet ice-edge, N.W.T. in 1986. Pp. 39-54 in Port and Ocean Engineering Under Arctic Conditions, Volume II, W.M. Sackinger et al., eds. Geophysics Institute, University of Alaska, Fairbanks, 111 pp.

Cox, T.M., A.J. Reed, A. Solow, and N. Tregenza. 2001. Will harbour porpoises (*Phocoena phocoena*) habituate to pingers? Journal of Cetacean Research Management 3:81-86.

Crum, L.A. 1995. Unresolved issues in bubble-related ambient noise. Pp. 243-269 in Sea surface Sound '94. M.J. Buckingham and J.R. Potter, eds. World Scientific, Singapore.

Cummings, W.C., and P.O. Thompson. 1971. Underwater sounds from the blue whale (*Balaenoptera musculus*). Journal of the Acoustical Society of America 50:1193-1198.

Cummings, W.C., and P.O. Thompson. 1994. Characteristics and seasons of blue and finback whale sounds along the U.S. West Coast as recorded at SOSUS stations. Journal of the Acoustical Society of America 95:2853.

Curtis K.R., B.M. Howe, and J.A Mercer. 1999. Low-frequency ambient sound in the North Pacific: Long time series observations. Journal of the Acoustical Society of America 106:3189-3200.

D'Amico, A., and W. Verboom, eds. 1998. Summary Record and Report, SACLANTCEN Bioacoustics Panel, La Spezia, Italy, 15-17 June, 1998, SACLANTCEN M-133, SACLANT Undersea Research Centre, 128 pp.

Dahlheim, M.E. 1987. Bio-acoustics of the gray whale (*Eschrichtius robustus*). Ph.D. Thesis, University of British Columbia, Vancouver, 315 pp.

Deane, G.B. 1999. Report on the Office of Naval Research Ambient Noise Focus Workshop, 9-11 August 1998, MPL Technical Report 463, Marine Physical Lab, Scripps Institution of Oceanography, 94 pp.

DeLuca, M. 2000. Seismic survey fleet shrinkage parallels industry contraction. Offshore Magazine, March, pp. 40-47.

Demski, L., G.W. Gerald, and A.N. Popper. 1973. Central and peripheral mechanisms in teleost sound production. American Zoology 13:1141-1167.

Diachok, O.I., and R.S. Winokur. 1974. Spatial variability of underwater ambient noise at the Arctic ice-water boundary. Journal of the Acoustical Society of America 55:750-753.

Diegez, C., M.D. Page, and M.F. Scanlon. 1988. Growth hormone neuroregulation and its alterations in disease states. Clinical Endocrinology 28:10-143.

Dragoset, W.H. 1990. Air-gun array specs: A tutorial, Geophysics. The Leading Edge, January, pp. 24-32.

Dragoset, W. 2000. Introduction to air guns and air gun arrays. The Leading Edge, May, pp. 892-897.

D'Spain, G.L., W.S. Hodgkiss, and G.L. Edmonds. 1991. Energetics of the deep ocean's infrasonic sound field. Journal of the Acoustical Society of America 89:1134-1158.

D'Spain, G.L., L.P. Berger, W.A. Kuperman, and W.S. Hodgkiss. 1997. Summer night sounds by fish in shallow water. Pp. 379-384 in Proceedings of International Conference on Shallow-Water Acoustics (SWAC'97).

D'Spain, G.L., L.P. Berger, W.A. Kuperman, J.L. Stevens, and G.E. Baker. 2001. Normal mode composition of earthquake T phases. Pure and Applied Geophysics 158:475-512.

Dubrovsky, N.A., and S.V. Kosterin. 1993. Noise in the ocean caused by lightning strokes. Pp. 697-709 in Natural Physical Sources of Underwater Sound, B.R. Kerman, ed. Kluwer Academic Publishing, Dordrecht, Holland.

Duffy, J.E. 1996. Eusociality in a coral-reef shrimp. Nature 381:512-514.

Dyer, I. 1987. Speculations on the origin of low frequency Arctic ocean noise. Pp. 513-532 in Sea Surface Sound, B.R. Kerman, ed. Proceedings-NATO. Advanced Research Series Workshop, Kluwer Academic Publishers, Dordrecht, Holland.

Edds, P.L. 1988. Characteristics of finback (*Balaenoptera physalus*) vocalizations in the St. Lawrence Estuary Canada. Bioacoustics 1:131-150.

Ellison, W.T., C.W. Clark, and G.C. Bishop. 1987. Potential use of surface reverberation by bowhead whales (*Balaena mysticetus*) in under-ice navigation: Preliminary considerations. Report of the International Whaling Commission 37:329-332.

Emery, L., M. Bradley, and T. Hall. 2001. Data base description (DBD) for the historical temporal shipping data base (HITS), Version 4.0, PSI Tech. Report TRS-301, Planning Systems Inc., Slidell, LA, 35 pp.

Eng, R. 2001. New Acquisition Methods Yield 3D Seismic Improvements. Offshore Magazine, March, pp. 48-50.

Engås, A., S. Løkkeborg, E. Ona, and A.V. Soldal. 1996. Effects of seismic shooting on local abundance and catch rates of cod (*Gadus Morhua*) and haddock (*Melanogrammus aeglefinus*). Canadian Journal of Fisheries and Aquatic Sciences 53:2238-2249.

Enger, P.S. 1981. Frequency discrimination in teleosts—central or peripheral? Pp. 243-255 in Hearing and Sound Communication in Fishes, W.N. Tavolga et al., eds. Springer-Verlag, New York.

Epifanio, C.L., J.R. Potter, G.B. Deane, M.L. Readhead, and M.J. Buckingham. 1999. Imaging in the ocean with ambient noise: The ORB experiments. Journal of the Acoustical Society of America 106:3211-3225.

Erbe, C. 1997. The masking of beluga whale (*Delphinapterus leucas*) vocalisations by icebreaker noise. University of British Columbia, Vancouver, 164 pp.

Erbe, C. 2000. Detection of whale calls in noise: Performance comparison between a beluga whale, human listeners, and a neural network. Journal of the Acoustical Society of America 108:297-303.

Erbe, C. 2002. Underwater noise of whale-watching boats and potential effects on killer whales (*Orcinus orca*), based on an acoustic impact model. Marine Mammal Science 18:394-418.

Erbe, C., and D.M. Farmer. 1998. Masked hearing thresholds of a beluga whale (*Delphinapterus leucas*) in icebreaker noise. Deep-Sea Research Part II—Topical Studies in Oceanography 45:1373-1388.

Erbe, C., and D.M. Farmer. 2000. Zones of impact around icebreakers affecting beluga whales in the Beaufort Sea. Journal of the Acoustical Society of America 108:1332-1340.

Erbe, C., A.R. King, M. Yedlin, and D. Farmer. 1999. Computer models for masked hearing experiments with beluga whales (*Delphinapterus leucas*). Journal of the Acoustical Society of America 105:2967-2978.

Estalote, E. 1984. NORDA acoustic models and databases. Navy Ocean Research Development Activity, Technical Note 293.

Etter, D.M. 1996. Introduction to MatLab for Engineers and Scientists. Prentice-Hall, Englewood Cliffs, NJ.

Etter, P.C. 1996. Underwater Acoustic Modeling: Principles, Techniques, and Applications. 2nd Edition. E&FN Spon, London, 344 pp.

Etter, P.C. 2001. Recent advances in underwater acoustic modeling and simulation. Journal of Sound and Vibration 240:351-383.

Etter, P.C., R.M. Deffenbaugh, and R.S. Flum, Sr. 1984. A survey of underwater acoustic models and environmental-acoustic data banks. ASW System Program Office, ASWR-84-001.

Evans, D.I., and G.R. England. 2001. Joint interim report Bahamas marine mammal stranding event of 15-16 March 2000. National Oceanic and Atmospheric Administration. Available at: http://www.nmfs.noaa.gov/prot_res/PR2/Health_and_Stranding_Response_Program/Interim_Bahamas_Report.pdf.

Everest, F.A., R.W. Young, and M.W. Johnson. 1948. Acoustical characteristics of noise produced by snapping shrimp. Journal of the Acoustical Society of America 20:137-142.

Fay, R.R. 1988. Hearing in vertebrates: A psychophysics databook. Hill-Fay Associates, Winnetka, IL.

Fay, R.R., and A.N. Popper. 2000. Evolution of hearing in vertebrates: The inner ears and processing. Hearing Research 149:1-10.

Fay, F.H., B.P. Kelly, P.H. Gehnrigh, J.L. Sease, and A.A. Hoover. 1984. Modern populations, migrations, demography, tropics, and historical status of the Pacific walrus. Pp. 231-376 in Environmental Assessment of the Alaskan Continental Shelf: Final Reports of Principal Investigators, Volume 37. U.S. National Oceanic and Atmospheric Administration, Anchorage, AK.

Felizardo, F.C., and W.K. Melville. 1995. Correlations between ambient noise and the ocean surface wave field. Journal of Physical Oceanography 25:513-532.

Ferguson, B.G., and J.L. Cleary. 2001. In situ source level and source position estimates of biological transient signals produced by snapping shrimp in an underwater environment. Journal of the Acoustical Society of America 109:3031-3037.

Ferris, R.H. 1972. Comparison of measured and calculated normal-mode amplitude functions for acoustic waves in shallow water. Journal of the Acoustical Society of America 52:981-988.

Finley, K.J., G.W. Miller, R.A. Davis, and C.R. Greene. 1990. Reactions of belugas, (*Delphinapterus leucas*) and narwhals (*Monodon monoceros*) to ice-breaking ships in the Canadian high arctic. Canadian Bulletin of Fisheries and Aquatic Science 224:97-117.

Finneran, J.J., D.A.Carder, and S.H. Ridway. 2002. Low frequency acoustic pressure, velocity, and intensity thresholds in a bottlenose dolphin (*Tursiops truncatus*) and white whale (*Delphinapterus leucas*). Journal of the Acoustical Society of America 111:447-456.

Fish, M.P. 1964. Biological sources of sustained ambient sea noise. Pp. 175-194 in Marine Bio-acoustics, W.N. Tavolga, ed. Pergammon Press, New York.

Fish, M.P., and W. Mowbray. 1970. Sounds of Western North Atlantic Fishes. The Johns Hopkins Press, Baltimore, MD, 207 pp.

Fish, J.F., and C.W. Turl. 1976. Acoustic source levels of four species of small whales. NUC TP 547. NTIS AD-A037620. U.S. Naval Undersea Center, San Diego, CA, 14 pp.

Fish, J.F., and J.S. Vania. 1971. Killer whale (*Orcinus orca*) sounds repel white whales (*Delphinapterus leucas*). Fisheries Bulletin 69:531-535.

Fletcher, H. 1940. Auditory patterns. Review of Modern Physics 12:47-65.

Fletcher, S., B.J. Le Boeuf, D.P. Costa, P.L. Tyack, and S.B. Blackwell. 1996. Onboard acoustic recording from diving northern elephant seals. Journal of the Acoustical Society of America 100:2531-2539.

Frisk, G.V. 1994. Ocean and seabed acoustics: A theory of wave propagation. Prentice-Hall, Englewood Cliffs, NJ, 299 pp.

Gaskin, D.E. 1976. The evolution, zoogeography, and ecology of Cetacea. Oceanography and Marine Biology Annual Review 14:247-346.

Geraci, J.R., S.A. Testaverde, D.J. St. Aubin, and T.H. Loop. 1978. A mass stranding of the Atlantic white-sided dolphin (*Lagenorhynchus acutus*): A study into pathobiology and life history. Report of the Marine Mammal Commission, PB-289 361. Washington, DC, 155 pp.

Gilbert, P.W., and H. Kritzler. 1960. Experimental shark pens at the Lerner Marine Laboratory. Science 132:424.

Gisiner, R.C. 1998. Proceedings: Workshop on the Effects of Anthropogenic Noise in the Marine Environment. Office of Naval Research, 141 pp.

Goldman, J. 1974. A model of broadband ambient noise fluctuations due to shipping. Bell Telephone Labs Report OSTP-31 JG.

Goold, J.C. 2000. A diel pattern in the vocal activity of short-beaked common dolphins (*Delphinus delphis*). Marine Mammal Science 16(1):240-244.

Gordon, J.C.D. 1987. Sperm whale groups and social behaviour observed off Sri Lanka. Reports of the International Whaling Commission 37:205-217.

Gordon, J.C.D., R. Leaper, F.G. Hartley, and O. Chappell. 1992. Effects of whale watching vessels on the surface and underwater acoustic behaviour of sperm whales off Kaikoura, New Zealand. NZ Department of Conservation. Science & Research Series 52:64.

Graham, N.E., and H.F. Diaz. 2001. Evidence for intensification of North Pacific winter cyclones since 1948. Bulletin of the American Meteorological Society 82:1869-1893.

Gray, L.M., and D.S. Greeley. 1980. Source level model for propeller blade rate radiation for the world's merchant fleet. Journal of the Acoustical Society of America 67:516-522.

Greene, C.R. 1985. Characteristics of wateborne industrial noise, 1980-84. Pp. 197-253 in Behavior, disturbance responses and distribution of bowhead whales *Balaena mysticetus* in the eastern Beaufort Sea, 1980-84, W.J. Richardson, ed. LGL Ecological Research Associates, Inc. 306 pp.

Hamson, R.M., and R.A. Wagstaff. 1983. An ambient-noise model that includes coherent hydrophone summation for sonar system performance in shallow water. SACLANT ASW Research Center Report SR-70.

Harlow, H.J., F.G. Lindzey, W.D. Van Sickle, and W.A. Gern. 1992. Stress response of cougars to nonlethal pursuit by hunters. Canadian Journal of Zoology 70:136-139.

Harris, F.C. 1978. On the use of windows for harmonic analysis with the discrete Fourier transform. Pp. 51-83 in Proceedings of the IEEE, Volume 66.

Harris, C.M. 1991. Handbook of Acoustical Measurements and Noise Control. 3rd Edition. McGraw-Hill, New York.

Hart's E&P. 2002. Available at: http://www.eandpnet.com/ep/previous/0102/0102marine_tech.htm.

Hastings, M.C., A.N. Popper, J.J. Finneran, and P.J. Lanford. 1996. Effect of low frequency underwater sound on hair cells of the inner ear and lateral line of the teleost fish (*Astronotus ocellatus*). Journal of the Acoustical Society of America 99:1759-1766.

Hattingh, J., and D. Petty. 1992. Comparative physiological responses to stressors in animals. Comparative Biochemistry and Physiology A 101:113-116.

Hawkins, A.D., and A.D.F. Johnstone. 1978. The hearing of the Atlantic salmon (*Salmo salar*). Journal of Fish Biology 13:655-673.

Holt, S.A. 2002. Intra- and inter-day variability in sound production by red drum (*Sciaenidae*) at a spawning site. Bioacoustics 12:227-230.

Holt, M.M., B.L. Southall, D. Kastak, C.J. Reichmuth, and R.J. Schusterman. 2002. Aerial hearing thresholds in pinnipeds: A comparison of free-field and headphone thresholds. Abstract. Presented to 14th Biennial Conference on the Biology of Marine Mammals, November 28-December 5, 2001, p. 102.

Irvine, A.B., M.D. Scott, R.S. Wells, and H.J. Kaufmann. 1981. Movements and activities of the Atlantic bottlenose dolphin (*Tursiops truncates*) near Sarasota, Florida. Fisheries Bulletin 79:671-688.

Ivey, L.E. 1991. Underwater Electroacoustic Transducers; USRD Transducer USRD Transducer Catalogue, Underwater Sound Reference Detachment, Naval Research Laboratory, Orlando, FL.

Jacobs, D.W., and W.N. Tavolga. 1967. Acoustic intensity limens in the goldfish. Animal Behavior 15:324-335.

Janik, V.M. 2000. Source levels and the estimated active space of bottlenose dolphin (*Tursiops truncates*) whistles in the Moray Firth, Scotland. Journal of Comparative Physiology A 186:673-680.

Jarvick, E. 1980. Basic Structure of Evolution of the Vertebrates. Volume 1. Academic Press, New York.

Jeannette, R.L., E.L. Sander, and L.E. Pitts. 1978. The USI array noise model, Version I documentation. Underwater Systems, Inc., USI-APL-R-8.

Jefferson, T.A., and B.E. Curry. 1994. Review and evaluation of potential acoustic methods of reducing or eliminating marine mammal-fishery interactions. Report from the Marine Mammal Research Program, Texas A&M University, College Station, for the U.S. Marine Mammal Commission, NTIS PB95-100384, Washington, DC.

Jensen, F.B., W.A. Kuperman, M.B. Porter, and H. Schmidt. 1994. Computational Ocean Acoustics. American Institute of Physics, New York, 612 pp.

Johnson, M.W. 1948. Sound as a tool in marine ecology, from data on biological noises and the deep scattering layer. Journal of Marine Research 7(3):443-458.

Johnson, J.S. 2001. Final Overseas Environmental Impact Statement and Environmental Impact Statement for Surveillance Towed Array Sensor System Low Frequency Active (SURTASS LFA) Sonar, Volumes 1 and 2.

Johnson, S.R., J.J. Burns, C.I. Malme, and R.A. Davis. 1989. Synthesis of information on the effects of noise and disturbance on major haulout concentrations of Bering Sea pinnipeds. OCS Study MMS 88-0092. Report from LGL Alaska Research Associates, Inc., for U.S. Minerals Management Service, Anchorage, AK, NTIS PB89-191373.

Johnson, M.P., P.L. Tyack, W.M.X. Zimmer, P.J.O. Miller, and A. D'Amico. 2001. Acoustic vocalizations ad movement patterns of a tagged sperm whale (*Physeter macrocephalus*) during foraging dives. Abstract. Presented at the 14th Biennial Conference on the Biology of Marine Mammals, Vancouver, BC, Canada.

Jones, M.L., and S.L. Swartz. 1984. Demography and phenology of gray whales and evaluation of whale-watching activities in Laguna San Ignacio, Baja California Sur, Mexico. Pp. 309-374 in The Gray Whale (*Eschrichtius robustus*), M.L. Jones et al., eds. Academic Press, Orlando, FL.

Kamminga, C. 1988. Echolocation signal types of odontocetes. Pp. 9-22 in Animal Sonar Processes and Performance, P.E. Nachtigall and P.W.B. Moore, eds. Plenum Press, New York.

Kastak, D., and R.J. Schusterman. 1999. In-air and underwater hearing sensitivity of a northern elephant seal (*Mirounga angustirostris*). Canadian Journal of Zoology 77:1751-1758.

Kerman, B.R., ed. 1988. Sea surface sound: Natural mechanisms of surface generated noise in the ocean. In Proceedings NATO Advanced Research Series Workshop 1987, Kluwer Academic Publishers, Dordrecht, Holland, 639 pp.

Kerman, B.R., ed. 1993. Natural Physical Sources of Underwater Sound. Kluwer Academic Publishers, Dordrecht, Holland, 750 pp.

Ketten, D.R. 1984. Correlations of morphology with frequency for odontocete cochlea: Systematics and topology. Ph.D. dissertation, The Johns Hopkins University, Baltimore, MD, 335 pp.

Ketten, D.R. 1994. Functional analyses of whale ears: Adaptations for underwater hearing. Institute of Electrical and Electronics Engineers Proceedings in Underwater Acoustics 1:264-270.

Ketten, D.R., and D. Wartzok. 1990. Three-dimensional reconstructions of the dolphin cochlea. Pp. 81-105 in Sensory Abilities of Cetaceans: Laboratory and Field Evidence, J.A. Thomas and R.A. Kastelein, eds. Plenum Press, New York.

Ketten, D.R., J. Lien, and S. Todd. 1993. Blast injury in humpback whale ears: Evidence and implications. Journal of the Acoustical Society of America 94:1849-1850.

Knudsen, V.O., R.S. Alford, and J.W. Emling. 1948. Underwater ambient noise. Journal of Marine Research 7:410-429.

Konagaya, T. 1980. The sound field of Lake Biwa and the effects of construction sound on the behavior of fish. Bulletin of the Japanese Society of Scientific Fisheries 46:129-132.

Kritzler, H., and L. Wood. 1961. Provisional audiogram for the shark *Carcharhinus leuces*. Science 133:1480-1482.

Kuiken, T., U. Hofle, P.M. Bennett, C.R. Allchin, J.R. Baker, E.C. Appleby, C.H. Lockyer, M.J. Walton, and M.C. Sheldrick. 1993. Adrenocortical hyperplasia, disease and chlorinated hydrocarbons in the harbour porpoise (*Phocoena phocoena*). Marine Pollution Bulletin 26:440-446.

Kuperman, W.A., and F. Ingenito. 1980. Spatial correlation of surface generated noise in a stratified ocean. Journal of the Acoustic Society of America 67:1988-1996.

Lair, S., P. Béland, S. De Guise, and D. Martineau. 1997. Adrenal hyperplastic and degenerative changes in beluga whales. Journal of Wildlife Diseases 33:430-437.

Lammers, M.O., W.W.L. Au, and D.L. Herzing. 2003. The broadband social acoustic signaling behavior of spinner and spotted dolphins. Journal of the Acoustical Society of America. In press.

Landau, L.D., and E.M. Lifshitz. 1987. Course of Theoretical Physics, Volume 6: Fluid Mechanics. 2nd Edition. Pergamon Press, New York, pp. 336-343.

Lasky, M., and R. Colilla. 1974. FANM-I fast ambient noise model. Program documentation and user's guide. Ocean Data Systems, Inc.

Lawson, J. 2002. Available at http://www.okgeosurvey1.gov/level2/nuke.cat.html.

Leighton, T.G., ed. 1997. Natural Physical Processes Associated with Sea Surface Sound. University of Southampton Press, Southampton, UK.

Leis, J.M., H.P.A. Sweatman, and S.E. Reader. 1996. What the pelagic stages of coral reef fishes are doing out in blue water: Daytime field observations of larval behavioral capabilities. Marine and Freshwater Research 47:401-411.

Lesage, V., C. Barrette, M.C.S. Kingsley, and B. Sjare. 1999. The effect of vessel noise on the vocal behavior of belugas in the St. Lawrence River Estuary, Canada. Marine Mammal Science 15:65-84.

LGL and Greeneridge. 1986. Reactions of beluga whales and narwhals to ship traffic and ice-breaking along ice edges in the eastern Canadian High Arctic 1982-1984. Environmental Studies 37. Indian and Northern Affairs Canada, Ottawa, Ontario, 301 pp.

Liberman, C.M. 2001. Charles Liberman Presentation Notes. Presentation to NRC Committee on Potential Impacts of Ambient Noise in the Ocean on Marine Mammals: Functionally important structural change in temporary and permanent NIHL in terrestrial mammals, Woods Hole, MA, June 11.

Lloyd's Register—Fairplay Ltd. 2001. World Fleet Statistics. Lloyd's Maritime Information Services, Stamford, CT.

Lombarte, A., and A.N. Popper. 1994. Quantitative analyses of postembryonic hair cell addition in the otolithic endorgans of the inner ear of the European hake [*Merluccius merluccius (Gadiformes, Teleostei)*]. Journal of Comparative Neurology 345:419-428.

Loughrey, A.G. 1959. Preliminary investigations of the Atlantic walrus [*Odobenus rosmarus rosmarus (Linnaeus)*]. Canadian Wildlife Service Wildlife Management Bulletin Series 1, Number 14, 123 pp.

Loye, D.P., and D.A. Proudfoot. 1946. Underwater noise due to marine life. Journal of the Acoustical Society of America 18:446-449.

Lukas, I.J., C.A. Hess, and K.R. Osborne. 1980. DANES/ASEPS Version 4.1 FNOC User's Manual. Ocean Data Systems, Inc.

Madsen, P.T., and B. Møhl. 2000. Sperm whales (*Physeter catodon* L. 1758) do not react to sounds from detonators. Journal of the Acoustical Society of America 107:668-671.

Madsen, P.T., R. Payne, N.U. Kristiansen, M. Wahlberg, I. Kerr, and B. Møhl. 2002. Sperm whale clicks: Bimodal, pneumatic sound production at depth. Journal of Experimental Biology 205:1899-1906.

Mahler, J.I., F.J.M. Sullivan, and M. Moll. 1975. Statistical methodology for the estimation of noise due to shipping in small sectors and narrow bands. Technical Memo W273. Bolt, Beranek, and Newman, Inc., Washington, DC.

Makris, N.C., and I. Dyer. 1986. Environmental correlates of pack ice noise. Journal of the Acoustical Society of America 79:1434-1440.

Makris, N.C., and I. Dyer. 1991. Environmental correlates of Arctic ice-edge noise. Journal of the Acoustic Society of America 90:3288-3298.

Maksoud, J. 2001. Seismic contractors encourage rebound by adding value. Offshore Magazine, March, pp. 40-47.

Malme, C.I., P.R. Miles, C.W. Clark, P. Tyack, and J.E. Bird. 1983. Investigations on the potential effects of underwater noise from petroleum industry activities on migrating gray whale behavior. Report No. 5366 submitted to the Minerals Management Service, U.S. Department of the Interior, NTIS PB86-174174, Bolt, Beranek, and Newman, Washington, DC.

Malme, C.I., P.R. Miles, C.W. Clark, P. Tyack, and J.E. Bird. 1984. Investigations on the potential effects of underwater noise from petroleum industry activities on migrating gray whale behavior. Phase II: January 1984 migration. Report No. 5586 submitted to the Minerals Management Service, U.S. Department of the Interior, NTIS PB86-218377. Bolt, Beranek, and Newman, Washington, DC.

Malme, C.I., P.R. Miles, P. Tyack, C.W. Clark, and J.E. Bird. 1985. Investigation of the potential effects of underwater noise from petroleum industry activities on feeding humpback whale behavior. BBN Report 5851, OCS Study MMS 85-0019. Report from BBN Laboratories Inc., Cambridge, MA, for U.S. Minerals Management Service, NTIS PB86-218385. Bolt, Beranek, and Newman, Anchorage, AK.

Malme, C.I., P.R. Miles, G.W. Miller, W.J. Richardson, D.G. Roseneau, D.H. Thompson, and C.R. Greene, Jr. 1989. Analysis and ranking of the acoustic disturbance potential of petroleum industry activities and other sources of noise in the environment of marine mammals in Alaska. BBN Report 6945, OCS Study MMS 89-0006. Report from BBN Systems and Technological Corporation, Cambridge, MA, for U.S. Minerals Management Service, NTIS PB90-188673. Bolt, Beranek, and Newman, Anchorage, AK.

Maniwa, Y. 1971. Effects of vessel noise in purse seining. In Modern Fishing Gear of the World, H. Krisjonnson, ed. Fishing News (Books) Ltd., London, UK.

Mann, D.A., Z. Lu, and A.N. Popper. 1997. A clupeiform fish can detect ultrasound. Nature 389:341.

Marshall, N.B. 1962. The biology of sound-producing fishes. Symposium of the Zoological Society of London 7:45-60.

Marshall, N.B. 1967. Sound-producing mechanisms and the biology of deep-sea fishes. Pp. 123-133 in Marine Bio-Acoustics II, W.N. Tavolga, ed. Pergamon Press, Oxford, UK.

Massa Products Corporation. 2002. Available at: http://www.massa.com/underwater.htm.

Mate, B.R., and J.T. Harvey, eds. 1987. Acoustical deterrents in marine mammal conflicts with fisheries. Oregon State University Sea Grant College Program, ORESU-W-86-001. Corvallis, OR, 116 pp.

Maybaum, H.L. 1993. Responses of humpback whales to sonar sounds. Journal of the Acoustical Society of America 94:1848-1849.

Mazzuca, L.L. 2001. Potential Effects of Low Frequency Sound (LFS) from Commercial Vehicles on Large Whales. M.S. Thesis, University of Washington, 70 pp.

McCarthy, E.M. 2001. International regulation of transboundary pollutants: The emerging challenge of regulating ocean noise. Ocean and Coastal Law Journal 6:257-292.

McCarthy, E., and J.H. Miller. 2002. Is anthropogenic ambient noise in the ocean increasing? Journal of the Acoustical Society of America 112(part2):2262.

McCauley, R.D. 2001. Biological sea noise in northern Australia: Patterns of fish calling. Ph.D. thesis, Department of Marine Biology, James Cook University of North Queensland, Australia, 282 pp.

McCauley, R.D., and D.H. Cato. 2001. Patterns of fish calling in northern Australia. Philosophical Transactions Series A. The Royal Society of London, UK.

McCauley, R.D., J. Fewtrell, A.J. Duncan, C. Jenner, M.N. Jenner, J.D. Penrose, R.I.T. Prince, A. Adihyta, J. Murdoch, and K. McCabe. 2000. Marine seismic surveys: Analysis and propagation of air-gun signals; and effects of exposure on humpback whales, sea turtles, fishes and squid. Australian Petroleum Production Association, Canberra, Australia, 198 pp.

McCauley, R.D., J. Fewtrell, and A.N. Popper. 2003. High intensity anthropogenic sound damages fish ears. Journal of Acoustical Society of America 113:1-5.

McKeen, S.W. 2002. Notice of the continuing effect of the List of Fisheries. Federal Register 67(12):2410-2419.

Medwin, H., and C.S. Clay. 1998. Fundamentals of Acoustical Oceanography. Academic Press, New York.

Miksis, J.L., M.D. Grund, D.P. Nowacek, A.R. Solow, R.C. Connor, and P.L. Tyack. 2001. Cardiac responses to acoustic playback experiments in the captive bottlenose dolphin (*Tursiops truncatus*). Journal of Comparative Psychology A 115:227-232.

Miller, P.J.O., N. Biasson, A. Samuels, and P.L. Tyack. 2000. Whale songs lengthen in response to sonar. Nature 405:903.

Milne, A.R. 1967. Sound propagation and ambient noise under sea ice. Pp. 120-138 in Underwater Acoustics, Volume 2, V.M. Albers, ed. Plenum Press, New York.

Mitson, R.B., ed. 1995. ICES Cooperative Research Report No. 209. Underwater Noise of Research Vessels Review and Recommendations. International Council for the Exploration of the Sea, Copenhagen, Denmark.

Mizrock, B. 1995. Alterations in carbohydrate metabolism during stress: A review of the literature. American Journal of Medicine 98:75-84.

Møhl, B. 1981. Masking effects of noise; their distribution in time and space. Pp. 259-266 in Proceedings: The Question of Sound from Icebreaker Operations, N.M. Peterson, ed. Arctic Pilot Project, Calgary, Alberta.

Møhl, B., M. Wahlberg, P.T. Madsen, L.A. Miller, and A. Surlykke. 2000. Sperm whale clicks: Directionality and source level revisited. Journal of the Acoustical Society of America 107:638-648.

Moll, M., R.M. Zeskind, and F.J.M. Sullivan. 1977. Statistical measures of ambient noise: Algorithms, program, and predictions. Bolt, Beranek, and Newman, Inc., Washington, DC.

Moll, M., R.M. Zeskind, and W.L. Scott. 1979. An algorithm for beam noise prediction. Report 3653, Bolt, Beranek, and Newman, Inc., Washington, DC.

Moore, K.E., W.A. Watkins, and P.L. Tyack. 1993. Pattern similarity in shared codas from sperm whales (*Physeter catodon*). Marine Mammal Science 9:1-9.

Morton, A.B., and H.K. Symonds. 2002. Displacement of *Orcinus orca* (L.) by high amplitude sound in British Columbia. ICES Journal of Marine Science 59:71-80.

Moulton, J.M. 1960. Swimming sounds and the schooling of fishes. Biological Bulletin 119:210-223.

Moulton, J.M. 1963. Acoustic behaviour of fishes. Pp. 655-693 in Acoustic Behaviour of Animals, R.G. Busnel, ed. Elsevier, Amsterdam.

Myrberg, A.A., Jr. 1972. Using sound to influence the behaviour of free-ranging marine animals. Pp. 435-468 in Behaviour of Marine Animals, Volume 2, H.E. Winn and B.L. Olla, eds. Plenum Press, New York.

Myrberg, A.A. 1980. Hearing in damsel fishes: An analysis of signal detection among closely related species. Journal of Comparative Physiology A 140:135-144.

Myrberg, A.A., Jr. 1981. Sound communication and interception in fishes. Pp. 395-426 in Hearing and Sound Communication in Fishes, W. N. Tavolga et al. eds. Springer-Verlag, New York.

Myrberg, A.A., Jr., C.R. Gordon, and A.P. Klimley. 1976. Attraction of free ranging sharks by low frequency sound, with comments on its biological significance. Pp. 205-228 in Sound Reception in Fish, A. Schuijf and A.D. Hawkins, eds. Plenum Press, New York.

Myrick, A.C. Jr., E.R. Cassano, and C.W. Oliver. 1990. Potential for physical injury, other than hearing damage, to dolphins from seal bombs used in the yellowfin tuna purse-seine fishery: Results from open-water tests. Admin. Rep. LJ-90-08. U.S. National Marine Fisheries Service, La Jolla, CA, 28 pp.

National Marine Manufacturers Association. 2002. Available at: http://www.nmma.org.
National Research Council (NRC). 1994. Low-Frequency Sound and Marine Mammals: Current Knowledge and Research Needs. National Academy Press, Washington, DC, 75 pp.
National Research Council (NRC). 2000. Marine Mammals and Low-Frequency Sound. National Academy Press, Washington, DC, 146 pp.
National Research Council (NRC). 2001. Climate Change Science. National Academy Press, Washington, DC, 29 pp.
Natural Resources Defense Council (NRDC). 1999. Sounding the Depths: Supertankers, Sonar, and the Rise of Undersea Noise. Natural Resources Defense Council, 75 pp.
Norris, K.S., G.W. Harvey, L.A. Burzell, and D.K. Kartha. 1972. Sound production in the freshwater porpoise Sotalia cf. fluviatilis Gervais and Deville and Inia geoffrensis Blainville in the Rio Negro Brazil. Investigations on Cetacea 4:251-262.
Nystuen, J.A. 1986. Rainfall measurements using underwater ambient noise. Journal of the Acoustical Society of America 79:972-982.
Nystuen, J.A., and D.M. Farmer. 1987. The influence of wind on the underwater sound generated by light rain. Journal of the Acoustical Society of America 82:270-274.
Offutt, G.C. 1970. Acoustic stimulus perception by the American lobster, Homarus americanus (Decapoda). Experientia 26:1276-1278.
Olesiuk, P.F., L.M. Nichol, M.J. Sowden, and J.K.B. Ford. 2002. Effect of the sound generated by an acoustic harassment device on the relative abundance and distribution of harbor porpoises (Phocoena phocoena) in Retreat Passage, British Columbia. Marine Mammal Science 18:843-862.
Osborne, K.R. 1979. DANES—a directional ambient noise prediction model for FLENUMOCEANCEN. Ocean Data Systems, Inc.
Payne, R., and D. Webb. 1971. Orientation by means of long range acoustic signaling in baleen whales. Annals of the New York Academy of Sciences 188:110-141.
Penner, R.H., C.W. Turl, and W.W. Au. 1986. Target detection by the beluga using a surface-reflected path. Journal of the Acoustical Society of America 80:1842-1843.
Personal Watercraft Industry Association. 2002. Available at: http://www.pwia.org.
Pew Oceans Commission. 2002. Available at: http://www.pewoceans.org. Data provided in December.
Piggott, C.L. 1964. Ambient sea noise at low frequencies in shallow water of the Scotian Shelf. Journal of the Acoustical Society of America 36:2152-2163.
Popper, A.N. 1980. Scanning electron microscopic studies of the sacculus and lagena in several deep-sea fishes. American Journal of Anatomy 157:115-136.
Popper, A.N., and R.R. Fay. 1997. Evolution of the ear and hearing: Issues and questions. Brain, Behavior and Evolution 50:213-221.
Popper, A.N., and R.R. Fay. 1999. The auditory periphery in fishes. Pp. 43-100 in Comparative Hearing: Fish and Amphibians, R.R. Fay and A.N. Popper, eds. Springer-Verlag, New York.
Popper, A.N., M. Salmon, and K.W. Horch. 2001. Acoustic detection and communication by decapod crustaceans. Journal of Comparative Physiology A 187:83-89.
Potter, J.R., and E. Delory. 1998. Noise sources in the seas and the impact for those who live there. In Proceedings of Acoustics and Vibration Asia (AVA).
Priestley, M. 1989. Spectral Analysis and Time Series. Academic Press, San Diego, CA.
Rabin, D., P.W. Gold, A.N. Margioris, and G.P. Chrousos. 1988. Stress and reproduction: Physiologic and pathphyhsiologic interactions between the stress and reproductive axes. Pp. 377-387 in Mechanisms of Physical and Emotional Stress, G.P. Chrousos, et al., eds. Plenum Press, New York.
Readhead, M.L. 1997. Snapping shrimp noise near Gladstone, Queensland. Journal of the Acoustical Society of America 101:1718-1722.

Renaud, D.L., and A.N. Popper. 1975. Sound localization by the bottlenose porpoise (*Tursiops truncatus*). Journal of Experimental Biology 63:569-585.

Renner, W.W. 1986a. Ambient Noise Directionality Estimation System (ANDES) User's Guide. Science Applications International Corporation (SAIC) 86-1705.

Renner, W.W. 1986b. Ambient Noise Directionality Estimation System (ANDES) Technical Description. Science Applications International Corporation (SAIC) 86-1645.

Renner, W.W. 1988. Ambient Noise Directionality Estimation System (ANDES) II User's Guide (VAX 11/78X installations). Science Applications International Corporation (SAIC) 88-1567.

ReVelle, D.O. 2001. Global infrasonic monitoring of large meteoroids. Journal of the Acoustical Society of America 109:2371.

Richardson, W.J., and C.I. Malme. 1993. Man-made noise and behavioral responses. Pp. 631-700 in The Bowhead Whale, J.J. Burns et al., eds. Special Publication No. 2, Society for Marine Mammalogy. Lawrence, KS.

Richardson, W.J., and B. Würsig. 1997. Influences of man-made noise and other human activities on cetacean behavior. Marine and Freshwater Behavior and Physiology 29:13-209.

Richardson, W.J., C.R. Greene, C.I. Malme, and D.H. Thomson. 1995. Marine Mammals and Noise. Academic Press, San Diego, CA, 576 pp.

Richardson, W.J., R.A. Davis, C.R. Evans, D.K. Ljungblad, and P. Norton. 1987. Summer distribution of bowhead whales, *Balaena mysticetus*, relative to oil industry activities in the Canadian Beaufort Sea, 1980-1984. Arctic 40:93-104.

Richardson, W.J., C.R. Greene, Jr., W.R. Koski, and M.A. Smultea, with G. Cameron, C. Holdsworth, G. Miller, T. Woodley, and B. Würsig. 1991. Acoustic effects of oil production activities on bowhead and white whales visible during spring migration near Pt. Barrow, Alaska—1990 phase. OCS Study MMS 91-0037. Report from LGL Ltd., King City, Ontario, for U.S. Minerals Management Service, NTIS PB92-170430, Herndon, VA.

Richter, C.F. 1958. Elementary Seismology. W.H. Freeman, San Francisco, CA, 768 pp.

Ridgway, S.H., E.G. Wever, J.G. McCormick, J. Palin, and J.H. Anderson. 1969. Hearing in giant sea turtle (*Chelonia myads*). Proceedings of the National Academy of Sciences 64:884-890.

Ridgway, S.H., J.G. McCormick, and E.G. Wever. 1974. Surgical approach to the dolphin's ear. Journal of Experimental Pathology 188(3):265-276.

Robinson, E.R., and S.O. McConnell. 1983. Sensitivity of high frequency surface-generated noise to sonar and environmental parameters. Pp. 11-15 in Proceedings: MTS/IEEE Oceans 83 Conference.

Rogers, P.H., and M. Cox. 1988. Underwater sound as a biological stimulus. Pp. 131-149 in Sensory Biology of Aquatic Animals, J. Atema et al., eds. Springer-Verlag, New York.

Romano, T., M.J. Keogh, C. Schlundt, J. Finneran, and D.A. Carder. 2001. Approaches to understanding the effects of loud sound on marine mammal health. Abstract. Presented to Fourteenth Biennial Conference on the Biology of Marine Mammals, November/ December, Vancouver, BC, Canada.

Ross, D. 1976. Mechanics of Underwater Noise. Pergamon Press, New York, 375 pp.

Ross, D.G. 1993. On ocean underwater ambient noise. Acoustics Bulletin, January/February, pp. 5-8.

Sancho, G., A.R. Solow, and P.S. Lobel. 2000a. Environment influences on the diel timing of spawning in coral reef fishes. Marine Ecology Progress Series 206:193-212.

Sancho, G., C.W. Peterson, and P.S. Lobel. 2000b. Predator-prey relations at a spawning aggregation site of coral reef fishes. Marine Ecology Progress Series 206:275-288.

Schevill, W.E., and W.A. Watkins. 1971. Pulsed sounds of the porpoise (*Lagenorhynchus australis*). Breviora 366:1-10.

Schreiner, H.F., Jr. 1990. The RANDI-PE noise model. Pp. 576-577 in Proceedings: IEEE Oceans 90 Conference.

Selye, H. 1973. The evolution of the stress concept. American Scientist 61:692-699.

Shankey, K.W., and P.J. McCabe, eds. 1976. SEPM Special Publication 59:109-121.

Southern, S. 2000. Molecular analysis of stress activated proteins and genes in dolphins and whales: A new technique for monitoring environmental stress. Abstract. Presented to Joint Conference of the American Zoo Veterinarians and the International Association of Aquatic Animal Medicine. September, New Orleans, LA.

Southern, S., A. Allen, N. Kellar, and A. Dizon. 2001. Molecular analysis of stress activated proteins and genes in cetaceans: A new methodology for monitoring environmental stress impact. Abstract. Presented to the 14th Biennial Meeting of the Society for Marine Mammalogists, November 28-December 4.

Spiess, F.N., J. Northrop, and E.W. Werner. 1968. Location and enumeration of underwater explosions in the North Pacific. Journal of the Acoustical Society of America 43:640-641.

Sparrow, V.W. 2002. Review and status of sonic boom penetration into the ocean. Journal of the Acoustical Society of America 111:537-543.

St. Aubin, D.J., and J.R. Geraci. 1992. Thyroid hormone balance in beluga whales, (*Delphinapterus leucas*): Dynamics after capture and influence of thyrotopin. Canadian Journal of Veterinary Research 56:1-5.

Stafford, K.M., C.G. Fox, and D.S. Clark. 1998. Long-range acoustic detection and localization of blue whale calls in the northeast Pacific Ocean. Journal of the Acoustical Society of America 104:3616-3625.

Stewart, B.S., W.E. Evans, and F.T. Awbrey. 1982. Effects of man-made waterborne noise on behavior of belukha whales (*Delphinapterus leucas*) in Bristol Bay, Alaska. HSWRI Technical Report 82-145. Report from Hubbs/Sea World Research Institute, San Diego, CA, for the U.S. National Oceanic and Atmospheric Administration, Juneau, AK, 29 pp.

Stirling, I., W. Calvert, and H. Cleator. 1983. Underwater vocalization as a tool for studying the distribution and relative abundance of wintering pinnipeds in the high arctic. Arctic 38:262-274.

Tavolga, W.N., ed. 1964. Marine Bio-acoustics. Pergammon Press, New York, 413 pp.

Tavolga, W.N. 1971. Sound production and detection. Pp. 135-205 in Fish Physiology, Volume 5, W.S. Hoar et al., eds. Academic Press, New York.

Terhune, J.M., and K. Ronald. 1986. Distant and near-range functions of harp seal (*Phoca groenlandica*) underwater calls. Canadian Journal of Zoology 64:1065-1070.

Terhune, J.M., T.C. Addy, T.A.M. Jones, and H.R. Burton. 2001. Underwater calling rates of harp and Weddell seals as a function of hydrophone location. Polar Biology 24:144-146.

Thode, A.M., G.L. D'Spain, and W.A. Kuperman. 2000. Matched-field processing, geoacoustic inversion, and source signature recovery of blue whale vocalizations. Journal of the Acoustical Society of America 107:1286-1300.

Thomas, J.A., and C.W. Turl. 1990. Echolocation characteristics and range detection threshold of a false killer whale *Pseudorca crassidens*. Pp. 321-334 in Sensory Abilities of Cetaceans: Laboratory and Field Evidence, J.A. Thomas and R.A. Kastelein, eds. Plenum Press, New York.

Thompson, P.O., and W.A. Friedl. 1982. A long term study of low frequency sounds from several species of whales off Oahu, Hawaii, USA. Cetology 45:1-19.

Thomas, J.A., and V.B. Kuechle. 1982. Quantitative analysis of Weddell seal (*Leptonychotes weddelli*) underwater vocalizations at McMurdo Sound, Antarctica. Journal of the Acoustical Society of America 72:1730-1738.

Thompson, T.J., H.E. Winn, and P.J. Perkins. 1979. Mysticete sounds. Pp. 403-431 in Behavior of Marine Animals, H.E. Winn and B.L. Olla, eds. Perseus Publishing, Cambridge, MA.

Thomas, J.A., R.A. Kastelein, and F.T. Awbrey. 1990. Behavior and blood catecholamines of captive belugas during playbacks of noise from an oil drilling platform. Zoo Biology 9:393-402.

Thorne, P.D. 1986. Laboratory and marine measurements on the acoustic detection of sediment transport. Journal of the Acoustical Society of America 80:899-910.

Tilt, W.C. 1985. Whales and whalewatching in North America with special emphasis on the issue of harassment. Yale School of Forestry and Environmental Studies. New Haven, CT, 122 pp.

Todd, S., P. Stevick, J. Lien, F. Marques, and D. Ketten. 1996. Behavioural effects of exposure to underwater explosions in humpback whales (*Megaptera novaeangliae*). Canadian Journal of Zoology 74:1661-1672.

Tolimieri, N., O. Haine, J.C. Montgomery, and A. Jeffs. 2003. Ambient sound as a navigational cue for larval reef fish. Bioacoustics. In press.

Tolstoy, I., and C.S. Clay. 1987. Ocean Acoustics: Theory and Experiment in Underwater Sound. Acoustical Society of America, New York, 293 pp.

Turl, C.W., R.H. Penner, and W.W.L. Au. 1987. Comparison of target detection capabilities of the beluga and bottlenose dolphin. Journal of the Acoustical Society of America 82:1487-1491.

Tyack, P.L. 2000. Functional aspects of cetacean communication. Pp. 270-307 in Cetacean Societies: Field Studies of Dolphins and Whales, J. Mann et al., eds. University of Chicago Press, Chicago.

Tyack, P.L., and C.W. Clark. 1998. Quick-look report: Playback of low-frequency sound to gray whales migrating past the central California coast. Unpublished.

U.S. Maritime Administration, Waterborne Databank. 2002. Foreign Waterborne; U.S. Army Corps of Engineers, Domestic Waterborne. Available at: http://www.marad. dot.gov/Marad_Statistics/US%20Waterborne%20Commerce%202000.htm.

U.S. Maritime Administration. 1999. Available at: http://www.marad.dot.gov/publications/.

U.S. Department of the Navy. 2001. Final Overseas Environmental Impact Statement and Environmental Impact Statement for Surveillance Towed Array Sensor System Low Frequency Active (SURTASS LFA) Sonar. Volume I.

University of California, Division of War Research. 1946. Underwater Noise Caused by Snapping Shrimp. Navy Department, Bureau of Ships, Contract NObs-2074 (formerly OEMsr-30), April 1.

Urick, R.J. 1975. Principles of underwater sound. McGraw-Hill, New York, 384 pp.

Urick, R.J. 1984. Ambient Noise in the Sea. Naval Sea Systems Command, Washington, DC.

van Bergeijk, W.A. 1967. The evolution of vertebrate hearing. Pp. 1-49 in Contributions to Sensory Physiology, Volume II, W.D. Neff, ed. Academic Press, New York.

Versluis, M., B. Schmitz, A. von der Heydt, and D. Lohse. 2000. How snapping shrimp snap: Through cavitating bubbles. Science 289(5487):2114-2117.

Wagstaff, R.A. 1973. RANDI: Research ambient noise directionality model. Naval Undersea Center, Technical Publication, 349 pp.

Wales, S.C., and R.M. Heitmeyer. 2002. An ensemble source spectra model for merchant ship-radiated noise. Journal of the Acoustical Society of America 111:1211-1231.

Wang Ding, B. Würsig, and W. Evans. 1995. Comparisons of whistles among seven odontocete species. Pp. 299-323 in Sensory Systems of Aquatic Mammals, R.A. Kastelein et al., eds. De Spil Publishers, Woerden, Netherlands.

Wartzok, D., and D.R. Ketten. 1999. Marine Mammal Sensory Systems. Pp. 117-175 in Biology of Marine Mammals, J.E. Reynolds III and S. Rommel, eds. Smithsonian Institution Press, Washington, DC.

Wartzok, D., W.A. Watkins, B. Würsig, and C.I. Malme. 1989. Movements and behaviors of bowhead whales in response to repeated exposures to noises associated with industrial activities in the Beaufort Sea. Report from Purdue University for Amoco Production Company, Anchorage, AK, 228 pp.

Wartzok, D., R. Elsner, H. Stone, B.P. Kelly, and R.W. Davis. 1992. Under-ice movements and the sensory basis of hole-finding by ringed and Weddell seals. Canadian Journal of Zoology 70:1712-1722.

Watkins, W.A. 1977. Acoustic behavior of sperm whales. Oceanus 20(2):50-58.

Watkins, W.A. 1980. Acoustics and behavior of sperm whales. Pp. 283-290 in Animal Sonar Systems, R.-G. Busnel and J.F. Fish, eds. Plenum Press, New York.

Watkins, W.A. 1981. The activities and underwater sounds of fin whales. Scientific Reports of the Whales Research Institute 33:83-117.

Watkins, W.A. 1986. Whale reactions to human activities in Cape Cod waters. Marine Mammal Science 2:251-262.

Watkins, W.A., and W.E. Schevill. 1974. Listening to Hawaiian spinner porpoises (Stenella cf. longirostris) with a three-dimensional hydrophone array. Journal of Mammalogy 55:319-328.

Watkins, W.A., and W.E. Schevill. 1975. Sperm whales (Physeter catodon) react to pingers. Deep-Sea Research 22:123-129.

Watkins, W.A., and W.E. Schevill. 1977. Sperm whale codas. Journal of the Acoustical Society of America 62:1485-1490 + phono record.

Watkins, W.A., and D. Wartzok. 1985. Sensory biophysics of marine mammals. Marine Mammal Science 1:219-260.

Watkins, W.A., K.E. Moore, and P. Tyack. 1985. Sperm whale acoustic behaviors in the southeast Caribbean. Cetology 49:1-15.

Webster, E.L., I.J. Elenkov, and G.P. Chousos. 1977. The role of cortictropin-releasing hormone in neuroendocrine immune interactions. Molecular Psychology 2:368-372.

Weinberg, H., and R.E. Keenan. 1996. Gaussian ray bundles for modeling high-frequency propagation loss under shallow-water conditions. Journal of Acoustical Society of America 100:1421-1431.

Weinberg, H., R.L. Deavenport, E.H. McCarthy, and C.M. Anderson. 2001. Comprehensive Acoustic System Simulator (CASS) Reference Guide. NUWC-NPT TM 01-016, Naval Undersea Warfare Center Division, Newport, RI.

Wenz, G.M. 1962. Acoustic ambient noise in the ocean: Spectra and sources. Journal of the Acoustical Society of America 34:1936-1956.

Wenz, G.M. 1964. Curious noises and the sonic environment in the ocean. Pp. 101-119 in Marine Bio-Acoustics, W.N. Tavolga, ed. Pergammon Press, New York.

Wenz, G.M. 1969. Low-frequency deep-water ambient noise along the Pacific Coast of the United States. U.S. Navy Journal of Underwater Acoustics 19:423-444, recently declassified.

Westwood, J., B. Parsons, and W. Rowley. 2002. Global Ocean Markets. Douglas-Westwood Associates, Canterbury, UK, 10 pp. Available at: www.dw-1.com.

Wilson, J.H. 1979. Very low frequency (VLF) wind-generated noise produced by turbulent pressure fluctuations in the atmosphere near the ocean surface. Journal of the Acoustical Society of America 66:1499-1507.

Wilson, O.B., Jr., S.N. Wolf, and F. Ingenito. 1985. Measurements of acoustic ambient noise in shallow water due to breaking surf. Journal of the Acoustic Society of America 78:190-195.

Winn, H.E. 1964. The biological significance of fish sounds. Pp. 213-230 in Marine Bio-acoustics, W.N. Tavolga, ed. Pergammon Press, New York.

Wood, F.G., and W.E. Evans. 1980. Adaptiveness and ecology of echolocation in toothed whales. Pp. 381-425 in Animal Sonar Systems, R.-G. Busnel and J.F. Fish, eds. Plenum Press, New York.

Yost, W.A., and M.C. Killion. 1997. Hearing thresholds. Pp. 1545-1554 in Encyclopedia of Acoustics, M.J. Crocker, ed. Wiley, New York.

Zakarauskas, P. 1986. Ambient noise in shallow water: A survey of the unclassified literature. Defense Research Establishment Atlantic Technical Memo 86/207, Defense Research Establishment Atlantic, Canada, 33 pp.

Zelick, R., D. Mann, and A.N. Popper. 1999. Acoustic communication in fishes and frogs. Pp. 363-411 in Comparative Hearing: Fish and Amphibians, R.R. Fay and A.N. Popper, eds. Springer-Verlag, New York.

Appendixes

Appendix
A

Committee and Staff Biographies

COMMITTEE CHAIR

George Frisk is a senior scientist in the Ocean Acoustics Laboratory in the Department of Applied Ocean Physics and Engineering (AOPE) at the Woods Hole Oceanographic Institution. From 1992 to 1997, he was chair of the AOPE department. He is the author of a textbook on ocean and seabed acoustics and has supervised or cosupervised 18 graduate students in the MIT/WHOI Joint Graduate Program in Applied Ocean Science and Engineering. His research interests include acoustic propagation, reflection, and scattering in the ocean and seabed, acoustic surface waves, scattering theory of waves, computational physics, inverse methods, seismoacoustic ambient noise, and Arctic acoustics and are funded primarily through the Office of Naval Research. Dr. Frisk is a fellow of the Acoustical Society of America and a member of the Institute of Electrical and Electronics Engineers and Sigma Xi.

COMMITTEE MEMBERS

David Bradley is a senior research scientist at the Applied Research Laboratory at Penn State University. His research is focused on environmental acoustics; in particular, he conducts research in acoustic radiation, propagation, scattering, reflection, absorption, and natural/man-made noise analysis, including spatial and temporal fluctuations. His research is funded in part by the Office of Naval Research. Dr. Bradley is a former member of the Ocean Studies Board and chaired the steering committee for the Sixth Symposium on Tactical Oceanography.

Jack Caldwell is a geophysical manager with WesternGeco in Houston, Texas. Dr. Caldwell is a leader in the geophysical services industry with regard to environmental issues related to marine seismic surveys. He was a member of the Minerals Management Service's High Energy Seismic Survey team, through which he has worked with numerous government agencies, the oil and gas industry, and the environmental communities on issues related to marine mammals and sound from industry seismic sources.

Gerald D'Spain is an associate research geophysicist at the Marine Physical Laboratory, Scripps Institution of Oceanography. Among Dr. D'Spain's research interests are ambient noise in the ocean and biological sounds, including marine mammal vocalizations and fish choruses and synthetic apertures in the ocean. His work is largely funded by the Office of Naval Research.

Jonathan Gordon is an honorary lecturer at the University of St. Andrews. His research has focused on cetacean acoustics, in particular, communication and echolocation in sperm whales, developing practical passive acoustic techniques in cetacean conservation (e.g., censusing, measuring length, assessing behavior), and studying and helping to mitigate the effects of man-made noise in marine mammals. Recent work includes studies on the effects of oceanographic and seismic sources on cetaceans and working with Birmingham Research University and Shell UK to develop more effective mitigation procedures for seismic surveys based on passive acoustic monitoring.

Mardi Hastings resigned from the committee in 2002 to accept a position at the Office of Naval Research. Prior to this position, she was an associate professor of mechanical engineering and biomedical engineering at the Ohio State University. Her major research interests include acoustics and vibrations, guided wave transmission, marine bioacoustics, and ultrasonics. Dr. Hastings is a registered professional engineer and member of several professional organizations, including the Acoustical Society of America, American Society of Mechanical Engineers, and Institute of Noise Control Engineering. She received the Presidential Young Investigator Award from the National Science Foundation and the Lumley Research Award from Ohio State and is listed in *Who's Who in Science and Engineering*. She was elected to be a fellow of the ASA in 1996.

Darlene Ketten holds a joint appointment as a senior scientist in biology at the Woods Hole Oceanographic Institution and as an assistant professor of otolaryngology at Harvard Medical School. Her research focuses on the sensory mechanisms of marine organisms, three-dimensional imaging and modeling of structural adaptations of aquatic vertebrates, underwater acoustics, underwater hearing, and diagnostic radiology of trauma and diseases of auditory systems. She receives research funding from the National Institutes of Health, the Office of Naval Research, the Seaver Foundation, and the National Marine Fisheries Service. Dr. Ketten is a fellow of the Acoustical Society of America.

James Miller is a professor in the Department of Ocean Engineering at the University of Rhode Island. Dr. Miller's research interests lie in underwater acoustics, tomography, and sonar. He has been at the forefront in the application of acoustic tomography imaging techniques in coastal waters and has studied the effects of shallow water on the performance of sonars. Recently, Dr. Miller has been particularly interested in mapping ocean-bottom properties using nonlinear tomographic techniques. He is investigating the use of high-frequency sonars for whale ship-strike avoidance. He has also investigated the effects of low-frequency underwater sound on marine mammals and divers.

Daniel L. Nelson is a senior program manager in the Physical Sciences Business Unit of BBN Technologies, in Cambridge, Massachusetts. He has more than 30 years of experience in marine acoustics research, development, and consulting including 15 years as the manager of BBN's Marine Acoustics and Mechanical Systems Department. Dr. Nelson is responsible for formulating approaches to solving complex marine acoustics problems, structuring the efforts necessary to implement these approaches, and overseeing all technical and managerial aspects of project planning and implementation. His primary areas of expertise include the measurement, analysis, prediction, and control of the underwater-radiated noise of ships and boats.

Arthur N. Popper is a professor in the Department of Biology and is director of the Neuroscience and Cognitive Science Program at the University of Maryland. His research focuses on understanding the basic structure and function of the auditory system in vertebrates, with particular interest in the ear of fishes and their sensory hair cells. Dr. Popper served on the Ocean Studies Board Committee on Low-Frequency Sound and Marine Mammals and was chair of the Committee to Review Results of the Acoustic Thermometry of the Ocean Climate's Marine Mammal Research Program. Dr. Popper is a fellow of the American Association for the Advancement of Science and the Acoustical Society of America. His research is funded by the National Institutes of Health, and he is the coeditor of the

Springer Handbook of Auditory Research, a series of over 20 books on the hearing sciences.

Douglas Wartzok is the vice-provost for academic affairs and dean of the University Graduate School of Florida International University. Dr. Wartzok served as the associate vice-chancellor for research, dean of the graduate school, and professor of biology at the University of Missouri-St. Louis for 10 years. For the past 30 years, his research has focused on sensory systems of marine mammals and the development of new techniques to study these animals and their use of sensory systems in their natural environment. He and his colleagues have developed acoustic tracking systems for studying seals and radio and satellite tracking systems for studying whales. For eight years he edited *Marine Mammal Science* and is now editor emeritus.

STAFF

Jennifer Merrill, *Study Director,* received her Ph.D. in marine and estuarine environmental science from the University of Maryland in 1999. A former NOAA Sea Grant Dean John A. Knauss Marine Policy fellow, she is now a program officer with the Ocean Studies Board of the National Research Council. In addition to this study, she directs a study of the feasibility of a coordinated international ocean exploration program and managed a workshop examining the future of marine biotechnology in the United States.

Julie Pulley, *Project Assistant,* received her B.S. in biology from Howard University in 1999. She has been with the Ocean Studies Board since 2001.

Appendix
B

Acronym List

ABR	auditory brainstem response
AHD	acoustic harassment device
AIM	Acoustic Integration Model
ANDES	Ambient Noise Directionality Estimation System
ANSI	American National Standards Institute
ATOC	Acoustic Thermometry of Ocean Climate
CASS	Comprehensive Acoustic System Simulation
CB	critical bandwidth
CNSR	critical noise-to-signal ratio
CR	critical ratio
CRF	corticotrophin releasing factor
DANM	Dynamic Ambient Noise Module
DAPS	Dynamic Ambient Noise Prediction System
ESME	Effects of Sound on the Marine Environment
FAO	Food and Agricultural Organization
FFT	fast Fourier transform
GRAB	Gaussian Ray Bundles
HITS	Historical Temporal Shipping

IOS	International Organization for Standardization
LFA	Low-Frequency Active (sonar)
MMPA	Marine Mammal Protection Act
NAS	National Academy of Sciences
NAVOCEANO	Naval Oceanographic Office
NMFS	National Marine Fisheries Service
NOAA	National Oceanic and Atmospheric Administration
NOPP	National Ocean Partnership Program
NRC	National Research Council
NRDC	Natural Resources Defense Council
NSF	National Science Foundation
OAML	Navy's Oceanographic and Atmospheric Master Library
ONR	Office of Naval Research
OSB	Ocean Studies Board
PE	parabolic equation
PTS	permanent threshold shift
RMS	root mean squared
RRF	range reduction factor
SOFAR	sound fixing and ranging
SOLMAR	Sound, Oceanography, and Living Marine Resources Database
SOSUS	Sound Surveillance System
SPAWAR	Space and Naval Warfare
TTS	temporary threshold shift
USFWS	U.S. Fish and Wildlife Service
USGS	U.S. Geological Survey

Appendix
C
Examples of Underwater Acoustics Noise Models

NAVY MODELS OF UNDERWATER NOISE

Ambient Noise

Ambient Noise Directionality System (ANDES; Renner, 1986a, b, 1988)
Ambient Noise Model (AMBENT; Robinson and McConnell, 1983)
Noise Model (CNOIS; Estalote, 1984; Osborne, 1979)
Directional Ambient Noise (DANES; Lukas et al., 1980)
Directional Underwater Noise Estimation System (DUNES; Bannister et al., 1989)
Fast Ambient Noise Model (FANM; Cavanagh, 1974a, b; Lasky and Colilla, 1974)
Normal Mode Ambient Noise (Kuperman and Ingenito, 1980)
Research Ambient Noise Directionality Model (Wagstaff, 1973; Hamson and Wagstaff, 1983; Schreiner, 1990; Breeding, 1993)
CASS/GRAB (Weinberg and Keenan, 1996; Weinberg et al., 2001)
Dynamic Ambient Noise Prediction System (DAPS)

BEAM NOISE STATISTICS

Analytic

BBN Shipping Noise (BBN Technologies Noise, Inc.; Mahler et al., 1975; Moll et al., 1977, 1979)
Bell Telephone Laboratories (BTL; Goldman, 1974)

USI Array Noise (Jeannette et al., 1978)
Sonobuoy Noise (Shankey and McCabe, 1976)

Simulation

Beam Program Library (BEAMPL; Etter et al., 1984)
Discreet Shipping Beam Noise Model (DSBN; Cavanagh, 1978)
Narrow Beam Towed Array Model (NABTAM; Etter et al., 1984)

Data compiled from P.C. Etter, 1996, Table 7.1.

Appendix
D
Research Recommendations from Previous NRC Reports (1994, 2000)

RECOMMENDATIONS—NRC (1994)
LOW-FREQUENCY SOUND AND MARINE MAMMALS:
CURRENT KNOWLEDGE AND RESEARCH NEEDS

Limitations of Current Knowledge

Data on the effects of low-frequency sounds on marine mammals are scarce. Although we do have some knowledge about the behavior and reactions of certain marine mammals in response to sound, as well as about the hearing capabilities of a few species, the data are extremely limited and cannot constitute the basis for informed prediction or evaluation of the effects of intense low-frequency sounds on any marine species.

Changes in the Proposed Regulatory Structure

It is the committee's judgment that the regulatory system governing marine mammal "taking" by research actively discourages and delays the acquisition of scientific knowledge that would benefit conservation of marine mammals, their food sources, and their ecosystems. The committee thus proposes several alternatives for reducing unnecessary regulatory barriers and facilitating valuable research while maintaining all necessary protection for marine mammals.Continued progress has been made on many of the recommendations made in previous reports (NRC, 1994; 2000) but further progress is needed in order to obtain the base of data needed to fully understand the impact(s) of sound on marine mammals and other marine organisms.

Topics for Future Research

Aims of research should be:

To determine the normal behaviors of marine mammals in the wild and their behavioral responses to human-made acoustic signals.

To determine how marine mammals utilize natural sounds for communication and for maintaining their normal behavioral repertoire.

To determine the responses of free-ranging marine mammals to human-made acoustic stimuli, including repeated exposure to the same individuals. How is the use of natural sounds altered by the presence of human-made sounds?

To determine how different sound types and levels affect migration and other movement patterns of marine mammals.

To determine the responses of deep-diving marine mammals to low-frequency sounds whose characteristics (source level, frequency, bandwidth, duty cycle) duplicate or approximate those produced by acoustic oceanographers.

To determine the structure and capabilities of the auditory system in marine mammals.

To determine basic hearing capabilities of various species of marine mammals.

To determine hearing capabilities of larger marine mammals that are not amenable to laboratory study.

To determine audiometric data on multiple animals in order to understand intraspecific variance in hearing capabilities.

Determine sound-pressure levels that produce temporary and permanent hearing loss in marine mammals.

To determine morphology and sound conduction paths of the auditory system in various marine mammals.

To determine whether low-frequency sounds affect the behavior and physiology of organisms that serve as part of the food chain for marine mammals.

To develop tools that can enhance observation and data gathering regarding marine mammal behavior or that can protect the animals from intense human-made sounds.

To develop tags that can be used for long-term observations of marine mammals, including studies on physiological condition, location (in three dimensions), sound exposure levels, and acoustic behavior.

To develop means of using in-place acoustic monitoring devices to study marine mammal movement and behavior on an ocean basin scale and of following individuals or groups of animals for extended periods and distances.

To develop procedures for rapid determinations of hearing capabilities (and perhaps other physiological studies) on beached or ensnared marine mammals.

To investigate the possibility of protecting marine mammals from some of the adverse effects of intense, low-frequency sounds by capitalizing on any normal avoidance reactions these animals might have to certain sounds.

RECOMMENDATIONS—NRC (2000)
MARINE MAMMALS AND LOW-FREQUENCY SOUND

Future Research and Observations Priority Studies

Recommendations: The committee supports the recommendation of NRC (1994) that there is a need for planned experiments designed to relate the behavior of specific animals to the received level of sound to which they are being exposed. Very few studies have succeeded in this aim. Because studies of ocean acoustics and marine mammal behavior are very challenging, successful experiments will require a closer collaboration between biologists and acousticians than has been the case in the past for many field studies. Success will also require continued refinement of techniques for making acoustic and visual observations, such as methods for locating vocalizing marine mammals and development of tags that can monitor received levels at the tagged animal.

To move beyond requiring extensive study of each sound source and each area in which it may be operated, a coordinated plan should be developed to explore how sound characteristics affect the responses of a representative set of marine mammal species in several biological contexts (e.g., feeding, migrating, and breeding). Research should be focused on studies of representative species using standard signal types, measuring a standard set of biological parameters, based on hearing type (Ketten, 1994), taxonomic group, and behavioral ecology (at least one species per group; reprinted as Box 4-1 in this report). This could allow the development of mathematical models that predict the levels and types of noise that pose a risk of injury to marine mammals. Such models could be used to predict in multidimensional space where TTS is likely (a "TTS potential region") as a threshold of potential risk and to determine measures of behavioral disruption for different species groups. Observations should include both trained and wild animals. The results of such research could provide the necessary background for future environmental impact statements, regulations, and permitting processes.

The uncertainty in predictions of received sound levels hinders the application of models of marine mammal responses to sound and will require three complementary approaches: (1) development of better acoustic propagation models; (2) development of better observing systems to gather the data needed in models; and when the first two are not feasible,

(3) development of better systems to observe ambient sound in the ocean and transient noise pollution events. Any research that includes relatively loud sound sources should monitor sound levels around the source site to gather data to calibrate their acoustic propagation models.

Acoustic studies focused on topics other than marine mammals should try to keep sound sources away from marine mammal "hotspots," even if this complicates logistics, increases costs, and/or decreases the efficiency of the experiments.

Studies of wild marine mammals should include careful determination of their locations, coupled with improved sampling and modeling of acoustic propagation to estimate received sound levels accurately. Alternatively, acoustic data loggers could be mounted on individual animals to record (1) the sounds (and their levels) to which the animals are exposed; (2) their vertical and horizontal movements; and (3) the sounds produced by the animals, including physiological sounds such as breathing and heartbeats.

A central theme of this report is that the task of developing predictive models of acoustic conditions that would harm marine mammals could be simplified by partitioning research among a small number of species that are representative in their hearing capabilities and sensitivities of larger groups of marine mammals. Box 4.1 (this report) describes the priority species groups, signal characteristics, and biological response parameters that should be investigated.

Richardson et al. (1995) summarized studies of marine mammal responses to human-generated sounds, particularly those associated with oil exploration and shipping. Some of these studies reported a significant difference between levels of pulsed versus more continuous sounds required to evoke a response in whales. To evoke the same level of response in migrating gray whales, a pulsed air-gun sound required levels 50 dB higher than a diverse array of low-frequency continuous sources. This result is unexpected based on human hearing capabilities. How do marine mammals respond to signals with durations between the pulsed air-gun noise (pulses separated by 7 to 15 s) and more continuous sounds? Another important question is: How do marine mammals respond when the received level is the same from two sources at different distances? This would help to discriminate whether marine mammals generally respond to received level (as was the case in the Phase II LFA study), estimated range to source, the gradient of acoustic energy over distance, and/or other sound characteristics.

Response to Stranded Marine Mammals

Recommendations: The concept of Stranded Whale Auditory Test (SWAT) teams recommended in NRC (1994) and NRDC (1999) should be implemented by funding trained scientists and associating them with stranding networks. The Office of Naval Research (ONR) partially funded a small effort to support the activities of a SWAT team, but the hardware and field methods are not yet adequate for wide testing. The ONR program manager (R. Gisiner) estimates that a considerable, but not unreasonable, amount of hardware and software design and testing will be needed (about one to two years of effort) before a system capable of regular operation under the SWAT team approach is feasible. However, this activity should be expanded to at least two teams, one on the East Coast and one on the West Coast of the United States. The teams should be responsible for (1) necropsy of suspected/possible marine mammal victims of sound injury (to be able to show whether sound caused the injuries or deaths) and (2) testing of hearing on stranded or entangled live animals. There is a need to expand the pool of individuals capable of doing this kind of work and capable of relating ear anatomy to function. An immediate need is for funding a specialist in evoked potential audiometry to develop improved methods applicable to large whales. A post-doctoral fellowship might be the most economical way to achieve this goal. The National Marine Fisheries Service (NMFS) and/or ONR should include funding for such work in the next budget cycle. Alternative possibilities for studying hearing in animals that are not kept in captivity also should be explored, such as placing a tag with electrodes on the head of a free-swimming whale and playing sound to the animal in a quiet environment.

Multiagency Research Support

Recommendations: If government funding shortages and priorities continue to constrain budgets for marine mammal research in the foreseeable future, management of sound in the ocean should remain conservative (and should incorporate management of all sources of human-generated noise in the sea, including industrial sources), in the absence of required knowledge. If government regulators need better information on which to base decisions, they should take such steps as necessary to provide increased funding for marine mammal research and to improve the ways that needed research is identified, funded, and conducted. Acquiring better information is often complicated because the regulatory parts of agencies like NMFS and the Fish and Wildlife Service (FWS) are separated from research, and funded research may not necessarily match research needed by regulators. It is imperative that the research and regulatory arms of NMFS and FWS maintain good linkages within these two agencies and that priority is given to research needed by regulators in each agency. Government agencies with basic science missions (e.g., National Science Foundation [NSF] and National Institutes of Health [NIH]) should fund marine mammal research at the levels needed to answer fundamental questions about hearing anatomy and physiology. Mission agencies with responsibilities related to marine mammals (e.g., ONR, National Oceanic and Atmospheric Administration [NOAA], Minerals Management Service (MMS), U.S. Geological Survey [USGS]) should also fund basic research (notwithstanding ONR's limitations under the Mansfield Amendment), in the spirit of the recommendation of NRC (1992) that "federal agencies with marine-related missions find mechanisms to guarantee the continuing vitality of the underlying basic science on which they depend." Such research should receive the same level of peer review as other basic research and be competitive with such programs for funding. Because marine mammal research is quite expensive, multiagency funding may be necessary to spread the costs. Alternatively, multiple parts of the same agency may need to cooperate in order to provide sufficient funds.

Multidisciplinary Research Teams and Peer Review

Recommendation: Consideration should be given to establishing a multiinvestigator program to study the effects of sound on marine mammals, funded by consortia of government agencies, nongovernmental organizations, shipping, and hydrocarbon exploration and production industries. These consortia should include individuals, organizations, and companies in nations that share marine mammal stocks and sound-producing activities with the United States (e.g., Canada, Mexico, nations of the North Atlantic Treaty Organization). Such consortia could be initiated through a workshop to bring together the interested communities. The design and implementation of auditory research on marine mammals ideally should be an interdisciplinary enterprise. Valuable contributions can be made by physical acousticians on the choice of sound stimuli to be used, by electronics experts on the choice and calibration of transducers for presenting the stimuli, by marine biologists on the choice of species and/or the best season and location for testing, by psychoacousticians on the testing procedures, and by statisticians on initial design and eventual data analysis and presentation. Without collaboration among specialists within these various disciplines, there is a greater probability that expensive and time-consuming projects will contain errors that preclude an unambiguous interpretation of the results. These projects are sufficiently complex that one or two individuals cannot reasonably be expected to have the full range of knowledge necessary for success. The logistical difficulties, permitting issues, and expense of such research demand advanced planning in all these areas.

If such a research program is established, it should use a public Request for Proposal (RFP) process that results in proposals from more than one research team and is modeled after the peer-review processes used by NSF and NIH. Conversely, some research should continue to be funded through the less conservative ONR model, which provides program managers with greater latitude to fund more innovative science. A spectrum of funding styles is useful. The RFP should be well advertised to encourage ideas and proposals from a wide range of researchers and institutions (including foreign participants), rather than relying on a set of traditional investigators. The goal of the process should be to optimize the selection of hypotheses, methods, and design and to identify the best performer(s) (e.g., best track record in previous work) for the proposed work. It is to the advantage of the sponsors to implement programs of broad-based peer review for such proposals. Future research on marine mammals unquestionably would profit from a broad-based review of the plans developed by multidisciplinary teams and evaluated by a peer-review process that is objective and independent. Such a review should determine whether the investigative team did the following adequately:

- identified basic problem(s);
- established specific hypotheses to be tested, with appropriate methods for data reduction, data presentation, and statistical analysis;
- identified optimal experimental methods and test conditions (including geographic location of study); and
- evaluated the power of the proposed experimental design.

Because long research projects often need to adjust to experience gained in field programs and learning about what kinds of observations are practical and achievable, it is important to provide advice from an outside review team later in the life of a project.

Sponsors of research need to be aware that studies funded and led by one special interest are vulnerable to concerns about conflict of interest. For example, research on the effects of smoking funded by NIH is likely to be perceived to be more objective than research conducted by the tobacco industry. Concern for peer review, efficiency, and independence argues for having an agency such as NSF take the lead in managing an interagency research program on the effects of noise on marine mammals.

Agencies that fund such applied research should ensure that adequate funding for analysis and plans for peer review are in place before a research award is made. Analysis might be speeded by employing a larger team for analysis and involving this team in planning the observations to make them as easy as possible to analyze later. Although publication in peer-reviewed journals is the standard by which most research is judged, applied research output from projects like the Marine Mammals Research Project (MMRP) is not necessarily suitable for publication in available academic journals and the results may need to be used for regulatory decisions within a shorter amount of time than the normal journal paper cycle. Timely peer review of such studies might be better accomplished by conducting a mail and/or panel review of results by an independent group established specifically for this purpose.

Population-Level Audiograms

Recommendations: Federal agencies should sponsor studies on the hearing abilities of both free-swimming and stranded animals. Population-level audiograms of many individuals (such as are performed for humans; see Yost and Killion, 1997) are necessary for establishing the baseline of hearing capabilities and normal hearing loss in marine mammals, as also recommended in NRC (1994). Stranded animals should be assessed to determine if their hearing is "normal." Data are needed to provide comparisons that would allow an evaluation of how common hearing deficits may be among stranded animals. The development of population-level audiograms will require the perfection and wide use of auditory evoked potential techniques, to eliminate the need to train all tested animals. However, if the cost and techniques limit widespread auditory evoked potential measurements of captive animals, a good sample of multiple animals (different ages and both genders) of the same species should be tested.

National Captive Marine Mammal Research Facility

Recommendations: If the studies described in Chapter 3 and Box 5.1 (NRC, 2000) are of sufficient priority to reduce uncertainties in the regulation of human-generated sound in the ocean, federal agencies should establish a national facility for the study of marine mammal hearing and behavior. The committee believes that such a facility might be established at relatively little incremental cost by enhancement of an existing Navy facility.

The facility for captive marine mammal research would have animals for "hire" by investigators funded for peer-reviewed research. Offset funds would come from individual grants and researchers, but the funding base for such a facility should not be provided solely by such offsets. Allocation of space, animals, and facility resources should be determined by a broad-based review board on the basis of the quality and significance of the proposed research. An additional virtue of establishing a national captive marine mammal research facility is that the total number of marine mammals removed from the wild would be minimized. Investigators could apply for support for short- or long-term study of the animals at this facility, from the range of agencies funding marine mammal research, at costs that would not have to include long-term maintenance of the animals. Such a facility should include the capability to work with trained animals in the open ocean. The Navy's Marine Mammal Program facility in San Diego keeps marine mammals and already has trained animals and expertise in maintaining them. Its role potentially could be expanded to provide a more widely accessible national facility, including unclassified research. If such a facility is operated by the Navy, it will be necessary to ensure that research data are not restricted from publication. Establishment of a facility to promote field studies could also enable research recommended in this report, but such a facility would be more expensive and a lower priority than a national facility for research on trained, captive animals.

Regulatory Reform

Recommendations: Congress should change the Marine Mammal Protection Act (MMPA) and/or NOAA should change the implementing legislation of the MMPA to allow incidental take authorization based solely on negligible impact on the population. Research should be undertaken to allow the definition of Level A harassment to be related to the TTS produced in a species, when known. Level B harassment should be limited to meaningful disruption of biologically significant activities that could affect demographically important variables such as reproduction and longevity.

Comprehensive Monitoring and Regulation of Sound in the Ocean

Recommendations: Noise monitoring is important and acoustic hotspots should be identified. Fortunately, ambient noise data exist for a variety of locations, which could provide time series and baselines for additional monitoring. Existing data should be identified and made accessible through a single easy-to-access source. Like marine mammal research programs, funding for noise monitoring should be awarded based on responses to a request for proposals and careful evaluation of the costs and benefits of the proposed systems. The opening of the existing IUSS for whale research was important for demonstrating the power of bottom-mounted hydrophone arrays, but the IUSS may or may not provide the best system for the acoustic monitoring tasks envisioned here, given that it was designed for an entirely different purpose.

The first step in comprehensive monitoring and regulation of sound in the ocean should be to attempt to characterize the existing ambient sound field in the ocean and to characterize the sources that contribute to it. Monitoring of baseline sound levels should be carried out, particularly in critical habitats of acoustically sensitive or vulnerable species or in habitats critical to specific life stages, such as breeding and calving areas. Protection of marine mammals from subtle or long-term effects of harassment cannot be achieved through regulation of individual "takes." An alternative habitat-oriented approach is required to protect marine mammals from the cumulative impacts of noise pollution, chemical pollution, physical habitat loss, and fishing. Such an approach requires monitoring of the status of marine mammal populations along with the quality of critical habitats, including the acoustic quality. Account should be taken of the populations involved; it is sensible to protect more rigorously species that are more endangered (e.g., northern right whales, *Eubalaena glacialis*) than those that are less at risk. Basic research regarding what is significant about critical habitats and what factors have population-level effects—for example, food supply, water quality, and noise levels and characteristics—will prove much more effective for protecting marine mammals than merely attempting to regulate individual human activities that may potentially cause changes in the behavior of an individual marine mammal. NMFS regulations should encompass the entirety of noise pollution and other threats to marine mammals.

Appendix
E

Glossary of Terms

UNDERWATER ACOUSTICS TERMS

This glossary contains definitions and explanations for many of the terms used in this report. Most of these definitions are consistent with those in the American National Standards Institute's (1994) "Acoustical Terminology" and those in Harris (1991). The text below indicates where the definitions in this report differ somewhat from the standard definitions. The first part of this glossary is divided into the following sections in order to group together concepts on a similar topic: "Noise and Statistical Analysis," "Physics of Sound," "Spectral Analysis and the Frequency Domain," "Temporal Character of Man-made Sounds," and "A Few Specific Sources of Noise."

Decibel—a logarithmic measure of the relative amplitude of two quantities. The two quantities being compared must have the same units so that their ratio is unitless. In underwater acoustics, the standard unit of acoustic pressure is the micro Pascal (μPa), or one-millionth of a Pascal. Therefore, the amplitude of acoustic pressure is compared to 1 μPa so that the sound pressure level (*SPL*) is defined as

$$SPL = 20 * \log_{10}(A_p / 1 \ \mu Pa)$$

where A_p is the pressure amplitude determined in a specified way (e.g., peak amplitude, RMS amplitude). The units of *SPL* are dB re 1 μPa. An equivalent way of defining the *SPL* is in terms of the square of the pressure amplitude,

$$SPL = 10*\log_{10}(A_p*A_p/(1\ \mu Pa)^2)$$

The deci in decibel indicates that the logarithm to the base 10 of squared pressure is multiplied by a factor of 10. This factor of 10 applies to quantities that are second order in the acoustic variables; for example, are proportional to the square of pressure, squared particle velocity amplitude, or the product of pressure and particle velocity. Examples of such quantities are acoustic energy density, magnitude of vector acoustic intensity, and acoustic power (see the "Physics of Sound" section of the Glossary). A factor of 20 is used for quantities at first order in the acoustic variables–acoustic pressure and acoustic particle velocity amplitude, for example. With regard to particle velocity, the sound particle velocity amplitude level ($SPVL$) can be defined in terms of the particle velocity amplitude, A_v, as

$$SPVL = 20*\log_{10}(A_v/1\ m/s) = 10*\log_{10}[A_v*A_v/(1\ m/s)^2]$$

It has units of dB re 1 m/s. More care must be taken in dealing with particle velocity because of its vector nature and because the polarization of the motion typically is more complicated than that of simple rectilinear motion.

The original definition of the decibel was given in terms of intensity amplitude ratios. This original definition is repeated in some modern textbooks. However, as indicated above, the decibel now is used in a much broader way, as can be seen in the national and international acoustics standards adopted by the American National Standards Institute (ANSI) and the International Organization for Standardization (ISO). In fact, those textbooks that define the decibel in terms of intensity amplitude ratios often proceed to report quantities in units of dB re 1 μPa or dB re 1 μPa²; these reference values pertain to quantities of pressure and pressure squared, respectively, and are not the units of intensity amplitude (which are W/m²).

Calls for the elimination of the decibel sometimes are heard. The decibel is here to stay, not only because it is part of ANSI and ISO standards, but because it is a valuable way (among others) of reporting acoustical quantities. It was invented and popularized for good reasons by the early pioneers in acoustics. The major reasons for its continued usefulness are given in Chapter 1, such as the fact that sound levels can span a large range of values (large dynamic range) and human perception of loudness appears to be logarithmic in nature. A far better recommendation than the elimination of the decibel is to insist that its reference units always be reported clearly.

Acoustic Source Properties, Sound Field Properties, and Properties of the Fluid Medium

Quantities that measure properties of the sound field and those that measure properties of the fluid medium itself must be clearly distinguished. For example, specific acoustic impedance is a property of a received sound field, whereas characteristic acoustic impedance is a property of the medium (see the "Physics of Sound" section of the Glossary). Another example is sound speed; group speed and phase speed are properties of a sound field, whereas medium sound speed obviously is a property of the medium. Acoustic density is the perturbation of the fluid density from its ambient value by the presence of sound and so is a property of the sound field. In contrast, the fluid ambient density is the density of the medium in the absence of sound. The same relationship holds for acoustic pressure and hydrostatic pressure.

Likewise, clear distinctions must be made between the properties of an acoustic source and those of a received field. For example, the sound level at a receiver is reduced from the source level by the transmission loss between source and receiver. (This transmission loss is quite large over short distances at close range from point-like sources as a result of spherical spreading.) The character of a received signal is due not only to the source of the signal but also the medium through which the signal has traveled. The received level is directly measured, whereas source level must be derived for many types of sources. For controlled, man-made sources that intentionally transmit sound such as sonars and air-gun arrays, the source level in most cases is well known. However, to derive the source level for uncontrolled and naturally occurring sources using underwater acoustic measurements of the received field, the location of the source must be known or determined, and the propagation conditions from source to receiver location must be accurately modeled. This effort has been accomplished successfully in situations for naturally occurring discrete sources that can be modeled as simple points in space such as individual vocalizing animals. However, estimating the source levels of spatially diffuse naturally occurring sources—chorusing fish schools, snapping shrimp colonies, ocean surface breaking waves, oscillating bubble clouds—is a topic of research. In other cases, the propagation conditions from source to receiver are too complicated to model with reliable accuracy at present and are areas of modern-day research. Examples of such naturally occurring sources in this category are earthquakes, surf, and lightning.

Noise and Statistical Analysis

Ambient Noise—the noise associated with the background din emanating from a myriad of unidentified sources. Its distinguishing features are that it is due to multiple sources, individual sources are not identified (although the type of noise source—e.g., shipping, wind—may be known), and no one source dominates the received field. Ambient noise is not necessarily that from distant sources, as sometime stated, since the collection of breaking waves directly above a receiver are not "distant" nor is thermal agitation. In addition, ambient noise in this report does not imply naturally occurring since ocean traffic noise has long been considered part of the ambient. Finally, although ambient noise is continuously present (at varying levels), the individual sources contributing to this background din do not have to create sounds continuously in time. The collection of individual snapping sounds from a colony of snapping shrimp, the clicking from a pod of sperm whales, or the sounds of breaking waves from a surface distribution of whitecaps typically are considered contributors to the ambient field even though the individual signals are transient in time.

Ambient vs. Specific Sources, Stochastic (Random) vs. Deterministic—the distinction between what is part of the ambient noise field and what is considered to originate from specific sources is somewhat arbitrary. For example, distant ships that contribute to the ambient noise field can become part of the set of specific sources with additional information. However, in any measurement or modeling effort, perfect knowledge of the contributing sources, their source characteristics, or the environment can never be achieved. The distinction between ambient noise and that from specific sources has a direct impact on modeling and the interpretation of signal processing results. Since *ambient* indicates a collection of sources not specifically identified, this component is modeled as stochastic in nature. That is, the properties of an ensemble (or collection) are the relevant features; for example, the probability of getting a 6 on the roll of a die. On the other hand, sounds from an identified source typically are modeled deterministically—the properties of a given realization are relevant. For example, a deterministic approach to rolling a die would predict what face of the die will appear given the die's initial position and velocity, its elastic properties and those of the table on which it lands.

The proper interpretation of data analysis results also requires an understanding of the distinction between stochastic and deterministic processes. For example, the spectral density of a continuous-in-time, aperiodic, deterministic signal is interpreted as the signal's mean squared amplitude per frequency, whereas for a stochastic signal it is the variance of the signal on a per frequency basis.

Noisy, Loud—describe the perception of sound and are not a property of the sound field itself. Their proper interpretation requires a clear indication of the perceiver of the sound; loud to whom, noisy to what species of marine mammal, and so on. Note that nowhere in this report is a comparison presented of the sound levels of various airborne sources and underwater acoustic sources. Such a comparison tends to anthropomorphize the effects of underwater noise sources. The issue here is not whether a given ocean acoustic source is "as loud as a jet engine" to a human, but rather how loud it is to a given marine species underwater. (The committee did not deal with the issue of the impact of airborne sound on marine mammals in air, only on the potential impact of airborne sound once it coupled into the underwater acoustic field.) Sounds that humans find bothersome may not have a significant impact on some marine species; conversely, sounds we cannot perceive may have significant adverse impact on some species in the marine environment.

Ocean Noise—the underwater sound from all types of noise sources, including noise from specific identified sources as well as ambient noise. For the purposes of evaluating the potential effects of underwater sound on the marine environment, both ambient noise and the noise from identified sources must be considered.

Variance—a statistical quantity that measures the variation (spread) of a random variable about its average (mean) value. For example, if the individual acoustic pressure samples as a function of time are assumed to be realizations of an underlying random process, the mean squared pressure is an estimate of the variance of the random process. The mean (a statistical measure of central tendency) of acoustic pressure and of acoustic particle velocity is equal to zero, by their definitions as variations about an equilibrium state. For stochastic (random) processes, the spectrum is the variance as a function of frequency. Likewise, the spectral density is the variance on a per frequency basis (Bendat and Piersol, 1986), and integration across all frequencies is equal to the broadband spectral level and to the variance of the original time series.

Other statistical properties of the ambient noise field are studied, such as its temporal and spatial coherence (measures of the degree of relatedness of two signals separated in time and in space, respectively; for example, see references in Urick, 1984) because of their relevance to signal and array processing and because they help identify certain noise source and propagation properties. A closely related property to spatial coherence is the directionality of the field, both vertically and azimuthally. The field's directional properties may be quite relevant to its potential impact on marine mammal hearing. For example, a diffuse interfering sound field may have a greater impact on masking than a highly directional one.

Physics of Sound

The field of physics contains words that also are used in common, everyday language. Examples are *intensity, power, work,* and *energy.* In physics these words have very specific and well-established definitions. However, they are often misused in the underwater acoustics literature. The most prevalent error probably is the use of *intensity* to describe the mean square pressure. Another common mistake is to use *power* when referring to instantaneous squared pressure and to refer to the sum of squared pressure over time as *energy.* The descriptions below conform to the physics definition of the terms as they apply to the study of acoustics. They are presented in this report to help remove the confusion that surrounds this topic.

Acoustic Energy Density—the energy per unit volume in the sound field. Two types of mechanical energy density exist in an acoustic field, potential energy density and kinetic energy density. The potential energy density measures the ability of the deformed fluid (deformed by the presence of sound) to do *work.* The acoustic kinetic energy density measures the ability of the fluid to do work because of the fluid motion associated with the acoustic field. The mean square pressure is proportional to the average potential energy density. Therefore, the integral of the pressure squared over a time interval is simply the mean squared pressure multiplied by the duration of the time interval, or proportional to the average potential energy density multiplied by the duration of the time interval. This time integral is not equal to *energy,* as it is sometimes mistakenly called. Similarly, the acoustic kinetic energy density is proportional to the mean squared particle velocity amplitude. The standard units of energy are joules, so that acoustic energy density (either potential or kinetic, or the sum of the two) has decibel units of dB re 1 J/m^3. Both types of acoustic energy density and acoustic energy (obtained by summing the energy densities over a specified volume of fluid) are second order in acoustic field variables.

Acoustic Impedance—a measure of the resistance to acoustic motion. There are two types of impedance that measure significantly different properties. *Characteristic impedance* is a property of the fluid medium itself and is equal to the product of the fluid ambient density (mass per unit volume in the absence of sound) and the speed of sound. The second type is *specific acoustic impedance.* It is a property of the sound field at a given point in space and is equal to the ratio of the acoustic pressure amplitude to acoustic velocity amplitude. As in the discussion of the decibel, the particle motion in acoustic fields can be quite complicated so that care is required in dealing with specific acoustic impedance. For example, acoustic velocity at a given frequency can have a component that is in quadrature with the acoustic pressure as well as one in phase, so that the specific acoustic impedance can have both real and imaginary components. In a few special sound fields, such as a single plane wave in a homogeneous fluid, the specific acoustic impedance equals the characteristic impedance. However, this equivalence does not hold in general.

Acoustic Intensity—the flow of acoustic energy through a surface with unit area per unit time. It is equivalent to acoustic energy *flux density*. Acoustic intensity equals the product of the acoustic pressure with the acoustic particle velocity, and therefore it is second order in acoustic field variables. As with particle velocity, acoustic intensity is a vector quantity; it has both a magnitude and a direction. The direction is the direction of flow of acoustic energy and is perpendicular to the surface with unit area referred to in the first line of this definition. The magnitude of the time-averaged flux density is not proportional to the mean squared acoustic pressure except in a few special types of sound fields. Energy flux density in acoustic fields can be categorized into two types. The first type, called *active intensity*, is the net flux of energy. It measures the propagating part of the sound field, which is that part of the field that can transfer information from one location to another. The second type of energy flux density has a time average of zero. However, this flow, the *reactive intensity*, is equally important in that it allows interference patterns in acoustic fields to exist. The units of acoustic intensity are those of energy per unit time per unit area. The standard units of energy are joules. A joule per second is equal to one watt. Therefore, acoustic intensity has units of W/m^2. In those rare cases in the underwater acoustics literature where true vector acoustic intensity is discussed, the reference unit of intensity typically is 1 pW/m^2 so that the corresponding decibel units are dB re 1 pW/m^2.

Acoustic Particle Velocity—the velocity of the fluid itself associated with the presence of a sound field. Velocity has both a magnitude and a direction (i.e., it is a vector quantity); pressure has magnitude only (it is a scalar quantity). The units of particle velocity amplitude in the SI system of units are m/s. A common assumption is that the acoustic particle velocity is "rectilinear"; that is, the particle motion is back and forth along a linear path along the direction of propagation. This type of motion occurs only in specific wave fields, such as a single propagating plane wave in a whole space or at specific points in space in more general types of wave fields. In ocean acoustic propagation, the particle can be elliptical or circular, both in the prograde and retrograde directions, as well as rectilinear. Acoustic particle velocity, along with acoustic pressure is a quantity that is first order in the acoustic field variables.

Acoustic Power—the integral over a well-defined surface area of the component of active vector acoustic intensity perpendicular to the surface.

Since the value of power will depend upon the area and orientation of the surface, as well as its location in the medium, the characteristics of the surface over which the integration is performed must be clearly specified. An important exception occurs when the area is that of a simple closed surface, such as a sphere. If the region enclosed by this simple closed surface does not include an acoustic source, the time-averaged acoustic power equals zero. That is, as much acoustic energy flows into the sphere as out of it, on average. This result is true for other simple closed surface shapes such as a cube or cylinder. If a source of sound is contained within the closed surface, the acoustic power measured by integrating the intensity over the surface is equal to the acoustic power of the source itself, regardless of the dimensions of the enclosing region (assuming that absorption of sound within the enclosed region is negligible). Therefore, for simple closed surfaces, acoustic power is a property of the source(s) contained within the region and is not a property of the sound field itself. Power is the time rate of change of energy in a system, so that the acoustic power of a source is the rate at which the source puts acoustic energy into the fluid medium. The units of acoustic power are joules per second, or watts, so that its decibel units are dB re 1 pW. As an example, a 75-watt lightbulb consumes nearly 139 dB re 1 pW of electrical power. Power, like intensity, is second order in acoustic field variables.

Acoustic Pressure—the force per unit area exerted by the fluid due as a result of its deformed state in a sound field. This force per unit area is analogous to the force exerted by a stretched or compressed spring. Acoustic pressure variations are variations of fluid pressure about an equilibrium value. In underwater acoustics, the equilibrium pressure is determined by the weight of the overlying water column (in atmospheric acoustics it is the weight of the overlying column of air) in the earth's gravitational field. The units of pressure in the SI system of units are pascals (Pa). In underwater acoustics the standard reference is one-millionth of a Pascal, called a micropascal (1 µPa). Acoustic pressure, like acoustic particle velocity, is first order in the acoustic field variables. The other acoustic field quantities presented in this section—acoustic energy density, acoustic intensity, and acoustic power—are second-order field quantities.

Adiabatic Incompressibility—the change in pressure (acoustic pressure plus ambient pressure) required to cause a unit fractional change in the fluid density. It measures how much force is needed to cause a given change in fluid volume. The *adiabatic* part of the term signifies that during the change in pressure, no heat or other form of energy is able to enter or escape the fluid undergoing deformation. Every type of wave motion requires some kind of force that tends to restore equilibrium conditions. In acoustics this restoring force is provided by the "springiness" of the fluid.

The adiabatic incompressibility, also called the *bulk modulus*, is the quantitative measure of the "stiffness" of the springiness of a fluid. It is derived by the fluid ambient density multiplied by the square of the medium sound speed.

Fluid Ambient Density—a property of a fluid that is equal to its mass per unit volume in the absence of sound. The term *ambient* as used here signifies the fluid's equilibrium state and should be distinguished from its use in the term *ambient noise*.

Shock Wave—an acoustic wave where the amplitude of the field is so large that the linear approximation to the governing physics equations is no longer valid and where discontinuities in acoustic quantities such as pressure and particle velocity can occur.

Sound—mechanical waves in a fluid that cause fluid motion and changes in pressure—compressions and dilatations—about an equilibrium state. The deformations in an individual freely propagating plane wave have a specific relationship between the temporal and spatial scales of variation in the direction of propagation. This relationship is given by the speed of sound. In addition, sound waves usually travel through a fluid medium without resulting in a net transport of the fluid itself—an exception occurs when the amplitude of the field becomes so large that nonlinear terms in the governing equations become important, as in a shock wave. The basic physics of sound is based on fundamental conservation laws (conservation of mass, conservation of momentum, and conservation of energy). A commonly held view of acoustics is, "I know sound when I hear it." However, this statement is not true, for example, when the wind is "heard" blowing past one's ears as a result of the pressure fluctuations associated with wind turbulence. It is important to distinguish between the physical properties of sound itself and the perception of sound by humans and animals.

Spectral Analysis and the Frequency Domain

*Note that some of the terms below (*spectrum *and* spectral density*) are defined in a somewhat different way than is found in textbooks.*

Autospectrum, Autospectral Density —*see* **Spectrum;** *see* **Spectral Density**

Fast Fourier Transform (FFT)—a computationally efficient algorithm for performing the Fourier transform with digitized data. The FFT can be viewed as a bank of narrow bandpass filters adjacent in frequency. The output of each filter is the equivalent amplitude and phase of the narrow band of frequency components centered on that filter's center frequency that are contained in the time series. Computer codes that implement the fast Fourier transform are readily available. Window functions often are used to taper the ends of a segment of time series prior to performing the transform in order to reduce the possibility of the spectral levels in one frequency band contaminating the levels in a significantly different frequency band ("spectral leakage"; resulting from the spectral sidelobes of the window function; see Harris, 1978). In order to numerically calibrate the FFT output to obtain an auto spectrum or autospectral density, the square of the Fourier transform amplitudes is normalized by the following quantities:

Autospectrum: square of the sum of the window values over the FFT length (fftl):

$$\left[\sum_{i=1}^{fftl} W(i)\right]^2$$

Autospectral Density: sum of the squared individual window values times the data sampling frequency (f_s):

$$f_s \sum_{i=1}^{fftl} W^2(i)$$

If a rectangular window is used so that $W(i) = 1/fftl$ for all $i = 1$ to fftl, the normalization factor for the autospectrum is equal to one, and that for the autospectral density equals $f_s/fftl$, which equals the FFT frequency resolution (bin width). These normalization factors pertain to "two-sided" autospectra and autospectral densities (i.e., those with both positive and negative frequencies); for one-sided quantities that span only the nonnegative frequencies, the normalization factors are half those given above (Bendat and Piersol, 1986).

Fourier Transform—a mathematical transformation that converts data values as a function of time (time series) into values as a function of frequency. In effect, the Fourier transform of a recorded piece of music describes the frequencies and levels of the individual notes (and their phases) that were played in creating the music. The transform is linear in the sense that the Fourier transform of a sum of quantities is equal to the sum of their transforms. The original time series can be reconstructed exactly from the Fourier transform output by an inverse transform. For this reason, a signal in the time domain and its corresponding Fourier transform in the frequency domain are considered transform pairs. An analogous relationship exists between the space and *spatial frequency* domains. (The inverse of the spatial frequency in a given spatial dimension is proportional to the wavelength of the wavefield in that dimension, just as the inverse of temporal frequency is proportional to the period of the wavefield.)

There exist several theorems that relate the properties of signals in the time domain and their corresponding Fourier transforms. These theorems can have important applications to the topic of this report. As a possible example, consider the rise time of an acoustic signal, which may be an important metric for evaluating the potential impact of a given sound on marine animals. A related concept is the rate of change of the signal amplitude with time, given by its derivative with time. A theorem in Fourier analysis states that the transform amplitude of the derivative of a signal is proportional to frequency multiplied by the transform amplitude of the original signal. Therefore, the higher-frequency components of a signal have greater importance in determining its time rate of change than the lower-frequency components (and must be present for a rapid rise time to occur). This theorem allows the Fourier transform and spectrum of the time rate of change of a quantity to be determined directly from the Fourier transform of the quantity itself.

Frequency—rate at which a repetitive event occurs, measured in hertz (Hz), cycles per second (from Richardson et al., 1995).

Infrasonic—describing sound that is lower in frequency than the minimum audible to humans generally below 20 Hz. Some baleen whales produce infrasonic sound.

Octave—a continuous band of frequencies in which the highest frequency is twice that of the lowest frequency.

Octave Band Levels—the spectral level obtained by integrating the spectral density across the octave band of interest.

One-third-Octave and One-third-Octave Band Levels—a third of an octave is a continuous band of frequencies in which the highest frequency is the cube root of 2 ($2^{1/3}$) times that of the lowest frequency. A one-third-octave band about a center frequency of F_c ranges from $F_c/(2^{1/6})$ to $F_c*(2^{1/6})$. The nominal standard bandwidth for the way in which the mammalian ear processes sound is a third octave. A one-third-octave band level is the spectral level obtained by integrating the spectral density across the one-third-octave band of interest.

Spectral Density (Spectral Density Function)—the spectrum per unit frequency. It is defined mathematically in terms of a limit, but numerical estimates are based on normalizing by the FFT binwidth. The autospectral density is the spectral density for a single time series of a specified quantity (versus the cross-spectral density, involving two different time series, which is not discussed in this report). The spectral density is the most appropriate quantity to use with signals whose spectral content varies continuously with frequency in a relatively smooth way ("continuous spectra"; see Priestley, 1989) since the numerical estimates of the spectral density levels of very narrow band components (*lines*) are dependent upon the FFT length. The spectrum level in a given frequency band (the *"band spectral level"*) can be obtained by integrating the spectral density across that band. Because of the overall increase in ocean ambient noise levels with decreasing frequency, band spectral levels are particularly sensitive to the lower frequency limit of the integration. The mean squared amplitude (or variance for random processes) of the original time series over a given time interval equals the integral of the spectral density for that time period across the whole frequency band. The decibel unit for the pressure spectral density in underwater acoustics is dB re 1 μPa^2/Hz and those for the particle velocity spectral density are dB re 1 $(m/s)^2$/Hz.

Spectral Level and Spectral Density Level—As used in this report, *spectral level* refers to either the band spectral level across a specified frequency band or the spectral level at a given frequency for narrowband tones, and *spectral density level* is the level of the spectral density. The two are not synonymous, in contrast to the definition in ANSI (1994).

Spectrum—in general, the frequency (temporal or spatial) dependence of some quantity. In this report, the *spectrum* of acoustic field quantities also is used for the *band spectrum* (across a specified frequency band) for continuous-in-frequency spectra and the spectrum at specific frequencies for discrete spectra (*line spectra*) (Bendat and Piersol, 1986) and refers to the squared amplitude of the Fourier transform of a quantity at first order in the acoustic field variables (pressure, particle velocity) as a function of frequency. The spectrum, as opposed to the spectral density, is the appropriate frequency domain description for signals composed of discrete frequency components (e.g., periodic signals such as those composed of a set of tonals). In that case, the spectrum levels are independent of the FFT length as long as the length is sufficient to resolve all contributing components (and the signal's time series does not change in character with change in the Fourier transform length). In contrast, the spectrum level of continuous spectra varies with varying FFT length since the amount of signal energy contained in each frequency bin changes with the binwidth and so actually represents a band spectral level. In calculating the spectrum from the Fourier transform, the phase is discarded so that a time series cannot be reconstructed from its spectrum. The pressure spectrum often is called the *power* spectrum, but it is not a measure of acoustic power (nor of electrical power if the original time series is a voltage signal). The term *power spectrum* in almost all cases is a misnomer and should be avoided unless true power in the physics sense is being considered. The units of the spectrum are the square of the units of the time series; the pressure spectrum has units of pressure squared. In underwater acoustics the squared pressure is referenced to 1 μPa^2 so that its spectrum in decibel units is dB re 1 μPa^2. As determined by the definition of the decibel and the properties of the logarithm, the decibel units of the pressure spectrum are equivalent to dB re 1 μPa. Similarly, the acoustic particle velocity amplitude spectrum has decibel units of dB re 1 $(m/s)^2$, equivalent to dB re 1 m/s.

The spectrum and spectral density of quantities at second order in the acoustic field variables, such as energy density, acoustic intensity, and acoustic power, are defined in terms of the spectra and spectral densities of the acoustic quantities at first order. For example, the potential energy density spectrum is equal to the pressure spectrum normalized by twice the adiabatic incompressibility, the kinetic energy density spectrum equals half the fluid ambient density times the particle velocity amplitude spectrum, and the acoustic intensity spectrum equals the cross-spectrum between the pressure and particle velocity (D'Spain et al., 1991). The acoustic power spectrum is the acoustic intensity spectrum integrated over a specified surface area.

ultrasonic—having a frequency above the human ear's audibility limit of about 20,000 Hz used of waves and vibrations.

Temporal Character of Man-made Sounds

In Chapter 2, man-made sources were categorized according to the activity involved in creating the sound, for example, seismic surveying, shipping, sonar use. A second way of organizing the noise created by these sources is according to their frequency content, as is done in the introduction to Chapter 2. The following table presents a third way of grouping man-made sounds based on their temporal character. The table also contains a listing of common metrics for each of the four categories, followed by comments on some of them. Note that some metrics are not appropriate for certain classes of signals, as discussed below.

TABLE E-1 Man-made Sounds Grouped by Temporal Character

Temporal Character	Some Common Metrics	Examples of Man-made Sources
Transient[1]	*Time domain* time series 0-pk amplitude[2] pk-pk amplitude[2] rise time total duration[3] mean squared amplitude[4] RMS amplitude[4] squared amplitude summed over total duration[5] *Frequency Domain* spectral density or spectrum	explosions sonic booms[1]
Continuous in time, Periodic[6]	*Frequency Domain* frequencies of tonals spectral levels of tonals spectrum[7] *Time Domain* maximum 0-pk amplitude maximum pk-pk amplitude mean squared amplitude rms amplitude	discrete tone sonars (commonly used in research) ships: propeller cavitation tonals prop-driven aircraft blade tip tonals machinery and pumps: engine rotation tonals
Periodic transient	*Time Domain* duty cycle period all those under Transient *Frequency Domain* repetition rate spectral density or spectrum of each transient	sonars (commercial, military, research) seismic air-guns and arrays pile driving pingers and AHDs
Continuous in time, Aperiodic	*Time Domain* mean squared amplitude rms amplitude 0-pk amplitude pk-pk amplitude *Frequency Domain* spectral density[7]	broadband ship cavitation dredging icebreaking

TABLE E-1 Continued

[1]Transient can signify transient in time (of short duration) or transient in space (passing through a certain region in a short period of time), or both. For example, a moving source like a surface ship or a supersonic aircraft radiates sound continuously in time but creates transient-in-time signals recorded by a fixed receiver. Therefore, it is important to specify whether transient applies to the source characteristics or to the received field.

[2]The zero-to-peak and peak-to-peak amplitudes are illustrated for an air-gun array signal in the upper panel of Figure 2-4.

[3]The total time duration of a signal emitted by a source usually can be defined unambiguously. However, when it pertains to a received signal, the total duration can depend on the level of ocean noise with respect to the received signal level ("signal-to-noise ratio") since noise can cover up the lower-level parts of the signal. In addition, propagation through the ocean can cause a change in the signal duration since the speed at which sound energy travels ("group velocity") can be a function of frequency, a phenomenon called "dispersion." Therefore, it is important to indicate clearly whether the signal duration is the duration emitted by the source or that measured at the receiver. This same point applies to all the other metrics in the table (see the discussion on source level versus received level in the Glossary).

[4]The time interval for the calculation of mean squared amplitude (the average of the squared amplitudes over a specified time interval) and root mean squared (RMS) amplitude (the square root of the mean squared amplitude) for a transient signal must be clearly specified in order for the quantity to be properly interpreted.

[5]Squared pressure integrated over the total signal duration is not equal to energy, as often stated. This issue is discussed in the "Physics of Sound" section. Rather, a more appropriate term might be "unweighted sound exposure." According to ANSI (1994), the term *sound exposure* is the "time integral of squared instantaneous frequency-weighted sound pressure over a stated time interval or event." The A-frequency weighting appropriate for human hearing sensitivity usually is used. However, no frequency weighing equivalent to unity weighting across the whole frequency band needs to be applied. Alternatively, a species-specific metric could be defined using a frequency weighting based on an audiogram (such as those plotted in Figure 1-1).

[6]When the term *continuous* is used, a clear distinction must be made between *continuous in time* and *continuous in frequency*. A continuous-in-time, periodic signal has a discrete spectrum, whereas a signal with a continuous-in-frequency spectrum can be either continuous-in-time and aperiodic or a transient-in-time domain.

[7]*Spectrum* and *spectral density* are defined in the "Spectral Analysis and the Frequency Domain" section of the Glossary. Note that a spectral density is not appropriate for sounds composed of a discrete set of tones (*line spectra*), since the spectral density level of a tone depends upon the FFT length. Conversely, the spectrum level of signals whose frequency content varies continuously with frequency ("continuous spectra") also varies with FFT length, since the FFT length determines the bandwidth over which the signal energy is integrated (see the "Spectral Analysis and the Frequency Domain" section of the Glossary). In these cases, the FFT bandwidth must be reported since these levels actually are "band," where the band is the FFT binwidth.

A Few Specific Sources of Noise

Cavitation—the tearing apart of a fluid when the negative pressure (dilatation) becomes sufficiently large. This process causes the formation of bubbles and the radiation of sound (Urick, 1975). Cavitation imposes an upper limit to the maximum acoustic power output of a sonar system. For example, at 3 kHz at shallow depths, Urick indicates that the cavitation threshold is slightly more than 1 atm = 1.013 bar = $1.013 \times 10^{11} \mu Pa$ = 220 dB re 1 μPa. Some cavitation can be tolerated so that the maximum levels can be a factor of 2 to 3 greater than the threshold, suggesting a maximum level of slightly more than 230 dB re 1 μPa. One reason for constructing arrays of sources is to create higher equivalent source levels along the array main beam in the far field than could be achieved by a compact source because of the limitations imposed by cavitation.

Microseisms—naturally occurring noise created by the nonlinear interaction of oppositely propagating ocean surface waves. Oppositely propagating waves give rise to a standing wave pattern that radiates sound with twice the frequency of that of the interacting surface waves. Microseisms are the dominant natural noise source in the space- and time-averaged ocean noise spectra below 5-10 Hz. Seismologists created the term *microseisms* because they also are the dominant source of noise in high-quality, on-land seismometer measurements; however, their source mechanism is unrelated to seismic processes in the solid earth. The Wenz curves (Plate 1) list "Seismic Background" above "Surface Waves—Second-Order Pressure Effects," but it is now known that the latter are the dominant source of prevailing ocean noise. Earthquakes and other tectonic processes contribute only intermittently.

Sonic Boom—a wave that is generated continuously by an object traveling faster than the speed of sound in the atmosphere. A sonic boom starts as a nonlinear shock wave with discontinuous jumps in pressure and fluid density. Because of dissipation and absorption, it eventually evolves into a linear acoustic wave at some distance from the source region. Its temporal character depends on the shape and size of the supersonic object, its speed, and its trajectory. The leading wavefront of a sonic boom is much like the bow wave of a surface ship, which is being "towed" along by the moving object. The sonic boom is a transient with respect to a receiver not traveling with the same velocity as the supersonic object creating the boom.

Thermal Noise—the pressure fluctuations associated with the thermal agitation of the ocean medium itself. It is what is left over when all other noise sources are removed and so provides the lowest bound for noise levels in the ocean. Thermal noise dictates the shape and level of ambient noise spectra above 50-100 kHz (depending on sea state; see Plate 1).

Appendix
F

Biological Terms

audiogram—graph showing an animal's absolute auditory threshold (threshold in the absence of much background noise) versus frequency. **Behavioral audiograms** are determined by tests with trained animals. Cf. evoked potential.*

A-weighting—a frequency response characteristic with the same sensitivity to frequency as that of the human ear. An A-weighted sound-level meter will have the same sensitivity (response) to sound at different frequencies as the average human ear.

baleen whale—whales in the order of Mysteceti that possess plates of dense, hair-like material (keratin) that hang side by side in rows from the roof of the mouth. These plates are for filter feeding on surface plankton and were formerly known as "whalebone" but have no actual resemblance to true bone.

beaked whale—members of the family *Ziphiidae*, which includes five genera: *Berardius*, *Hyperoodin*, *Mesoplodon*, *Tasmacetus*, and *Ziphius*.

catecholamine—any of various amines (as epinephrine, norepinephrine, and dopamine) that function as hormones or neurotransmitters or both.

cephalopod—a member of a group of mollusks including squids, cuttlefish, and octopuses.

cetacean—any member of the order Cetacea of aquatic, mostly marine mammals that includes whales, dolphins, porpoises, and related forms; among other attributes they have a long tail that ends in two traverse flukes.

critical band (CB)—frequency band within which background noise has strong effects on detection of a sound signal at a particular frequency.*

critical ratio (CR)—difference between sound level for a barely audible tone and the spectrum level of background noise at nearby frequencies.*

echolocation†—a physiological process for locating distant or invisible objects (as prey) by means of sound waves reflected back to the emitter (as a bat) by the objects.

ecosystem†—the complex of a community of organisms and its environment functioning as an ecological unit.

elasmobranchs†—any of a subclass (Elasmobranchii) of cartilaginous fishes that have five to seven lateral to ventral gill openings on each side and that comprise the sharks, rays, skates, and extinct related fishes.

glucocorticoid—steroids such as cortisol and corticosterone produced by the adrenal cortex and affecting a broad range of metabolic and immunologic processes.

habituation (behavioral)—gradual waning of behavioral responsiveness over time as animals learn that a repeated or ongoing stimulus lacks significant consequences for the animal (cf. sensitization).*

hair cell—a special kind of cell that has tiny hairs projecting from its surface into the intercellular space. Movement of the hairs is registered by neurons that contact the hair cell. Hair cells are found in the inner ear of mammals.

haulout—the act of a seal leaving the ocean and crawling onto land or ice.

homeostasis—a relatively stable state of equilibrium or a tendency toward such a state among the different but interdependent elements or groups of elements of an organism, population, or group.

hyperplasia—an abnormal or unusual increase in the elements composing a part (as cells composing a tissue).

hypertrophy—excessive development of an organ or part; specifically an

crease in bulk (as by thickening of muscle fibers) without multiplication
f parts.

hydrophone—transducer for detecting underwater sound pressures; an underwater microphone.*

invertebrate—lacking a spinal column; also of or relating to invertebrate
animals.

Level A harassment—any act of pursuit, torment, or annoyance that has the
potential to injure a marine mammal stock in the wild.

Level B harassment—any act of pursuit, torment, or annoyance that has the
potential to disturb a marine mammal or marine mammal stock in the wild
by causing disruption of behavioral patterns, including, but not limited to,
migration, breathing, nursing, breeding, feeding, or sheltering.

masking—obscuring of sounds of interest by interfering sounds, generally
at similar frequencies.*

mysticete—member of the suborder *Mysticeti,* the toothless or baleen
(whalebone) whales, including the rorquals, gray whales, and right whales;
the suborder of whales that includes those that bulk feed and cannot
echolocate. Their skulls have an antorbital process of maxilla, a loose
mandibular symphysis, a relatively small pterygoid sinus, and the maxillary
bone telescoped beneath the supraorbital process of the frontal, or baleen
whales, composed of four families: *Eschrichtiidae, Balaenidae,
Neobalaenidae,* and *Balaenoptidae.*

odontocete—member of the toothed-whale suborder *Odontoceti,* which
contains nine families and includes dolphins and porpoises: *Physeteridae,
Kogiidae, Monodontidae, Ziphiidae, Delphinidae, Pontoporiidae,
Platanistidae, Iniidae,* and *Phocoenidae.* The toothed whales, including
sperm and killer whales, belugas, narwhals dolphins, and porpoises; the
suborder of whales including those able to echolocate. Their skulls have
premaxillary foramina, a relatively large pterygoid sinus extending anteri-
orly around the nostril passage, and the maxillary bone telescoped over the
supraorbital process of the frontal.

Otarid—the eared seals (sea lions and fur seals), which use their foreflippers
for propulsion.

Phocid—a family group within the pinnipeds that includes all of the "tru seals (i.e., the "earless" species). Generally used to refer to all recei pinnipeds that are more closely related to *Phoca* than to *otariids* or th walrus.

pinniped—one of a group of acquatic, mostly marine, carnivorous animals; includes seals, sea lions, and walruses; all their limbs are finlike and they spend at least some time on land or ice.

Permanent Threshold Shift (PTS)—prolonged exposure to noise causing permanent hearing damage.

sensitization—an increased behavioral (or physiological) responsiveness occurring over time, as an animal learns that a repeated or ongoing stimulus has significant consequences. Cf. habitutation.

---

*Richardson et al., 1995.
†Webster.com